Fun Math Problem Solving For Elementary School

David Reynoso
John Lensmire
Kevin Wang

Kelly Ren

TITLES PUBLISHED BY ARETEEM INSTITUTE

Cracking the High School Math Competitions (and Solutions Manual) - Covering AMC 10 & 12, ARML, and ZIML

Mathematical Wisdom in Everyday Life (and Solutions Manual) - From Common Core to Math Competitions

Geometry Problem Solving for Middle School (and Solutions Manual) - From Common Core to Math Competitions

Fun Math Problem Solving For Elementary School (and Solutions Manual)

ZIML Math Competition Book Division E 2016-2017

ZIML Math Competition Book Division M 2016-2017

ZIML Math Competition Book Division H 2016-2017

ZIML Math Competition Book Jr Varsity 2016-2017

ZIML Math Competition Book Varsity Division 2016-2017

COMING SOON

Fun Math Problem Solving For Elementary School Vol. 2 (and Solutions Manual)

Counting & Probability for Middle School (and Solutions Manual) - From Common Core to Math Competitions

Number Theory Problem Solving for Middle School (and Solutions Manual) - From Common Core to Math Competitions

The books are available in paperback and Kindle eBook formats. To order the books, visit https://areteem.org/bookstore.

ISBN: 1-944863-07-9
ISBN-13: 978-1-944863-07-4

First printing, October 2017.

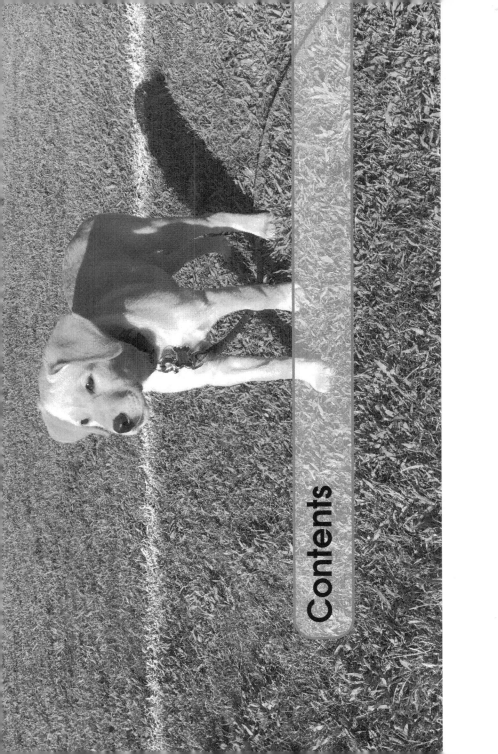

Contents

Introduction

This book is part of the ongoing effort by Areteem Institute to inspire students, parents, and teachers to gain a deeper understanding and appreciation of mathematics. This book is aimed for students in 3rd, 4th, and 5th grade in elementary school. This book leads readers through complex math concepts via age-appropriate approaches, such as fun stories in real-life scenarios, riddles and puzzles, magic tricks, cartoon drawings, jokes, etc. Math is fun! The authors of the book are experts in math who are passionate educators and they work hard to present the fun aspect of math to young students to stimulate interest in math and develop problem solving and critical thinking skills at an early age. In addition, this book reviews and expands state math standards, including the Common Core Standards, particularly the Operations and Algebraic Thinking (OA), Numbers and Operations in Base Ten (NBT), and Measurement and Data (MD) domains at the 3rd, 4th, and 5th grade level.

The book is divided into 8 chapters. In each of the chapters we introduce a new concept as well as step by step solutions to a variety of problems related to that particular concept. Each chapter contains 10 example questions with full solutions, 10 quick response questions and 25 practice problems. The problems are designed to test the students' mastery of the material discussed in each chapter.

Introduction

The book is available as a **Student Workbook** and has an accompanying Solutions Manual with full solutions. The Student Workbook contains all the material and practice problems, and answers to all practice problems. The Solutions Manual includes in-depth solutions to all of the quick response and practice problems.

The problems in this book offer the student a chance to start developing problem solving techniques that will be useful not only in mathematics but also in everyday life.

While most of the problems in the last few chapters of the book could be solved using algebraic techniques, we strongly encourage the students to attempt solving them without algebra. Not only the solutions are easier to understand, but students gain a deeper understanding of the problems this way, and the students develop logical thinking skills through creative reasoning. Some students may think that in solving these problems "we are doing the same as if we were solving equations", however, when solving equations we usually do not identify what the middle steps of the solving process mean for the problem itself, which may play a huge part into understanding basic abstract mathematical concepts.

Common Core and This Book

Teachers and students working in 3rd, 4th, and 5th grade math can use this book to teach and learn mathematical reasoning and problem solving, focusing on concepts in the OA (Operations and Algebraic Thinking), NBT (Numbers and Operations in Base Ten), and MD (Measurement and Data) Common Core domains.

For reference, a summary of these domains is provided below.

Operations and Algebraic Thinking

Standard(s)	Cluster
3.OA.1-4	Represent and solve problems involving multiplication and division.
3.OA.5-6	Understand properties of multiplication and the relationship between multiplication and division.
3.OA.7	Multiply and divide within 100.
3.OA.8-9	Solve problems involving the four operations, and identify and explain patterns in arithmetic.
4.OA.1-3	Use the four operations with whole numbers to solve problems.
4.OA.4	Gain familiarity with factors and multiples.
4.OA.5	Generate and analyze patterns.
5.OA.1-2	Write and interpret numerical expressions.
5.OA.3	Analyze patterns and Relationships.

Numbers and Operations in Base Ten

Standard(s)	Cluster
3.NBT.1-3	Use place value understanding and properties of operations to perform multi-digit arithmetic.
4.NBT.1-3	Generalize place value understanding for multi-digit whole numbers.
4.NBT.4-6	Use place value understanding and properties of operations to perform multi-digit arithmetic.
5.NBT.1-4	Understand the place value system.
5.NBT.5-7	Perform operations with multi-digit whole numbers and with decimals to hundredths.

Measurement and Data

Standard(s)	Cluster
3.MD.1-2	Solve problems involving measurement and estimation of intervals of time, liquid volumes, and masses of objects.
3.MD.3-4	Represent and interpret data.
3.MD.5-7	Geometric measurement: understand concepts of area and relate area to multiplication and to addition.
3.MD.8	Geometric measurement: recognize perimeter as an attribute of plane figures and distinguish between linear and area measures.
4.MD.1-3	Solve problems involving measurement and conversion of measurements from a larger unit to a smaller unit.
4.MD.4	Represent and interpret data.
4.MD.5-7	Geometric measurement: understand concepts of angle and measure angles.
5.MD.1	Convert like measurement units within a given measurement system.
5.MD.2	Represent and interpret data.
5.MD.3-5	Geometric measurement: understand concepts of volume and relate volume to multiplication and to addition.

The start of each chapter summarizes the specific Common Core standards emphasized in the chapter. In addition, the problem solving stressed in the exercises allows students to practice the other standards simultaneously, even if those standards are not the focus of the chapter.

For more details about the specific standards, clusters, and domains quoted above, see www.corestandards.org/Math where the full Mathematical Standards are available for download.

About Areteem Institute

Areteem Institute is an educational institution that develops and provides in-depth and advanced math and science programs for K-12 (Elementary School, Middle School, and High School) students and teachers. Areteem programs are accredited supplementary programs by the Western Association of Schools and Colleges (WASC). Students may attend the Areteem Institute through these options:

- Live and real-time face-to-face online classes with audio, video, interactive online whiteboard, and text chatting capabilities;
- Self-paced classes by watching the recordings of the live classes;
- Short video courses for trending math, science, technology, engineering, English, and social studies topics;
- Summer Intensive Camps on prestigious university campuses and Winter Boot Camps;
- Practice with selected daily problems for free, monthly ZIML competition at http://ziml.areteem.org.

The Areteem courses are designed and developed by educational experts and industry professionals to bring real world applications into STEM education. The programs are ideal for students who wish to build their mathematical strength in order to excel academically and eventually win in Math Competitions (AMC, AIME, USAMO, IMO, ARML, MathCounts, Math Olympiad, ZIML, and other math leagues and tournaments, etc.), Science Fairs (County Science Fairs, State Science Fairs, national programs like Intel Science and Engineering Fair, etc.) and Science Olympiad, or purely want to enrich their academic lives by taking more challenges and developing outstanding analytical, logical thinking and creative problem solving skills.

Since 2004 Areteem Institute has been teaching with methodology that is highly promoted by the new Common Core State Standards: stressing the conceptual level understanding of the math concepts, problem solving techniques, and solving problems with real world applications. With the guidance from experienced and passionate professors, students are motivated to explore concepts deeper by identifying an interesting problem, researching it, analyzing it, and using a critical thinking approach to come up with multiple solutions.

Thousands of math students who have been trained at Areteem achieved top honors and earned top awards in major national and international math competitions, including Gold Medalists in the International Math Olympiad (IMO), top winners and qualifiers at the USA Math Olympiad (USAMO/JMO), and AIME, top winners at the Zoom

International Math League (ZIML), and top winners at the MathCounts National. Many Areteem Alumni have graduated from high school and gone on to enter their dream colleges such as MIT, Cal Tech, Harvard, Stanford, Yale, Princeton, U Penn, Harvey Mudd College, UC Berkeley, UCLA, etc. Those who have graduated from colleges are now playing important roles in their fields of endeavor.

Further information about Areteem Institute, as well as updates and errata of this book, can be found online at http://www.areteem.org.

Acknowledgments

This book contains many years of collaborative work by the staff of Areteem Institute. This book could not have existed without their efforts. The materials in this book were prepared by Kelly Ren and Kevin Wang for Areteem's Young Math Olympians courses, and were later updated and expanded by John Lensmire and David Reynoso. Especially, the illustrations in each chapter were created by David. Yes, the same David Reynoso who is a mathematician by trade and artist by nature!

The examples and problems in this book were either created by the Areteem staff or adapted from various sources, including other books and online resources. We extend our gratitude to the original authors of all these resources.

Last but not least, special thanks go to Saber and Hazel, who starred in the pictures on the covers and chapter images, photographed by Kelly Ren.

1. It Matters Where It Is

Georgia's family lives in a house on a street with all 5-digit street numbers. The numbers of the houses on that street are all 200 apart. The first two digits of the house numbers are all 15. The last two digits are all 09. How many houses could be on that street?

Would you rather have \$10 or 10¢? Why are the numbers 13 and 31 different even though they each have one 1 and one 3? How do we immediately know that 5000 is bigger that 3999? We use place values everyday to tell numbers apart. In this chapter we'll practice using place values and solve a few riddles along the way!

The concepts introduced in this chapter directly correspond to Common Core Math Standards as shown in the following table.

3rd Grade	3.OA.3, 3.OA.4, 3.NBT.2
4th Grade	4.NBT.1, 4.NBT.2
5th Grade	5.MD.1

In addition to the standards above, problems and concepts in this section will help strengthen understanding of the following domains.

3rd Grade	3.OA, 3.NBT
4th Grade	4.OA, 4.NBT
5th Grade	5.OA, 5.NBT

1.1 Example Questions

Example 1.1

What 3-digit number has 8 hundreds, 1 more ten than hundreds, and no ones?

Solution

Since our number will have no ones, that means the ones digit is 0. We already know that the hundreds digit is 8 and the tens digit has to be $8 + 1 = 9$. So, the number we are looking for is 890.

Example 1.2

What 3-digit numbers have 2 hundreds and 2 more ones than tens?

Solution

We need to make sure that the first digit is 2, but it seems that for the other two digits we have some freedom. Say the second digit is 0, then the last digit has to be $0 + 2 = 2$ and we get the number 202. If we take a look at the number 213, its ones digit also is 2 more than the tens digit. Note that we can continue doing this depending on what digit we choose to have for the tens. So, when should we stop? We know that the biggest digit is 9, so if the ones digit is 9, the biggest tens digit we can have is 7, which would give us the number 279. So, all the numbers that we can write such that "have 2 hundreds and 2 more ones than tens" are 202, 213, 224, 235, 246, 257, 268 and 279.

Example 1.3

What is the difference between the largest 5-digit number and the smallest 5-digit number?

Solution

To find the largest possible 5-digit number, we want to make sure to use the biggest possible digit in each *place*. So, for the largest number we will use only 9s: 99999. The smallest is a little bit tricky. We know that the digit with the least value is 0, however, a 5-digit number cannot start with 0, so for the first digit we must use the second lowest digit we have available: 1. This means that the smallest number we can make with 5 digits is 10000. The difference is

$$99999 - 10000 = 89999$$

Example 1.4

Georgia's family lives in a house on a street with all 5-digit street numbers. The numbers of the houses on that street are all 15. The first two digits of the house numbers are all 200 apart. The last two digits are all 09. How many houses could be on that street?

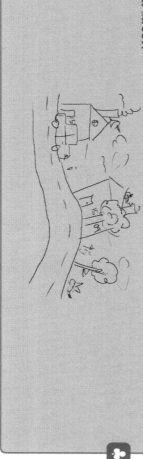

Solution

The smallest possible 5-digit number we can make that starts with 15 and ends with 09 is 15009. If that was the number of one of the houses, then since all the other house numbers are 200 apart from each other, the other house numbers would be 15209, 15409, 15609 and 15809. If we had that the first house number was 15109, we could still come up with other 4 house numbers that follow the rules: 15309, 15509, 15709 and 15909. In both cases, there are 5 houses.

Example 1.5

I'm thinking of a 2-digit number. I'll give you some clues to see if you can figure out what it is:

- The sum of the two digits is 8.
- If I switch the two digits, the new number is 18 more than the number I'm thinking of.

What number am I thinking of?

Solution

There are several ways that we could add two digits up to 8:

$$0 \& 8, \quad 1 \& 7, \quad 2 \& 6, \quad 3 \& 5, \quad \text{and} \quad 4 \& 4.$$

Note that we don't want to use 0, because if we switch the digits of a number that ends in 0, we get a one-digit number. So, our number should be one of 17, 71, 26, 62, 35, 53 or 44. We know that when we switch the numbers we get a bigger number than the original, that means that it should be either 17, 26 or 35 since the other ones become smaller (or stay the same) when we swap the digits. If we look at the differences between those numbers and their "switched" versions, we get: $71 - 17 = 57$, $62 - 26 = 36$ and $53 - 35 = 18$. So the number we are looking for is 35.

Example 1.6

Do you get an odd number or even number if you add 9 even numbers?

Solution

Remember, all even numbers end up in either 0, 2, 4, 6 or 8. Every time we add up two numbers that end with those digits, we will get again a number that ends up in either 0, 2, 4, 6 or 8. That means that *every time* we add up two even numbers we get again an even number. Therefore, if we add up 9 even numbers we will still get an even number.

Example 1.7

Find a 6-digit number that has no repeated digits and when multiplied by its last digit is equal to 999999.

Solution

The ones digit of our number has to be either 3 or 7, since those are the only 1-digit numbers that end in 9 when we square them. So the 6-digit number we are looking for will be either $999999 \div 3 = 333333$ or $999999 \div 7 = 142857$. Since we cannot repeat digits, the number we want is 142857.

Example 1.8

Leslie just broke her piggy bank and is trying to figure out how much money she has. She counted the number of coins of each kind that she has: 354 pennies, 44 nickels, 12 dimes, 72 quarters, 46 half dollars, and 5 silver dollars. How much money did Leslie have in her piggy bank?

Solution

Each of the coins has a different value, exactly like the values of the digits in a number depending on their place. We will multiply the number of coins of each kind by the *face value* of the coin and add everything together to figure out how much money she has in total.

$$354 \times 0.01 + 44 \times 0.05 + 12 \times 0.10 + 72 \times 0.25 + 46 \times 0.50 + 5 \times 1.00 = 52.94.$$

This is easier to see in a table:

Type of coin	Value of coin (in $)	# of coins	Total (in $)
Penny (1¢)	0.01	354	3.54
Nickel (5¢)	0.05	44	2.20
Dime (10¢)	0.10	12	1.20
Quarter (25¢)	0.25	72	18.00
Half dollar (50¢)	0.50	46	23.00
Silver dollar ($1)	1.00	5	5.00
Total			52.94

So, Leslie had $52.94 in her piggy bank.

Example 1.9

Troy bought some candies at the local grocery store in town. He had to pay $6.23, so he gave the cashier a $10 bill. The cashier gave him back three $1 bills and... 77 pennies! It seems there were no other coins available in the cash register. No one likes pennies and neither does Troy. He looked in his pockets for some coins he could use to pay for the 23 cents, and he found 1 dime, 3 quarters, 2 pennies and 2 nickels. What coins will help him get the smallest amount of pennies back?

Solution

If Troy used the dime, the two nickels, and the two pennies he would get

$$10 + 2 \times 5 + 2 \times 1 = 22$$

cents in total, but he needs at least 23¢ to avoid getting all those pennies back. By just using one quarter he gets 25¢, which is just a little bit more than the 23¢ he needs, so if he gives the cashier the $10 bill and one quarter, the cashier would have to give him back

$$10.25 - 6.23 = 4.02$$

dollars, that is, four $1 bills and two pennies.

Example 1.10

Randy has 18 coins. Some are pennies, some are nickles and some are quarters. He has two times as many quarters as pennies and three times as many nickles as quarters. How much money does Randy have?

Solution

If Randy had only 1 penny, he would have $2 \times 1 = 2$ quarters and $3 \times 2 = 6$ nickels, so he would have $1 + 2 + 6 = 9$ coins in total. He has twice as many coins, so he should have twice as many of each. That is, Randy has 2 pennies, 4 quarters and 12 nickels.

We know the face value of each of the coins, so we can find how much money he has in total:

$$2 \times 0.01 + 4 \times 0.25 + 12 \times 0.05 = 0.02 + 1.00 + 0.60 = 1.62 \text{ dollars.}$$

1.2 Quick Response Questions

Problem 1.1 What numbers are shown below?

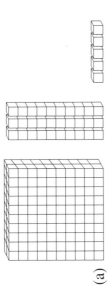

(a)

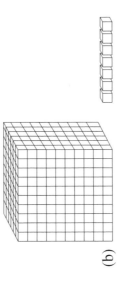

(b)

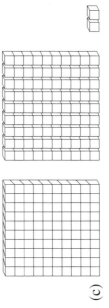

(c)

Problem 1.2 Do you get an odd number or even number if you add 10 odd numbers?

Problem 1.3 Can you find three numbers whose sum is 100 with only one of the numbers odd? Why or why not?

Problem 1.4 Do you get an odd or even number if you multiply an odd number by an even number?

Problem 1.5 Add any two odd numbers. The ones digit of the sum is always _____.

Problem 1.6 In 98960, which digit is the tens place?

Problem 1.7 What is the sum of the digits in the number one million?

Problem 1.8 My aunt owns many cars. If the number of cars my aunt owns is 1 less than the thousands digit of 17854, how many cars does she own?

Problem 1.9 Fill in the blanks

(a) 5000 = _____ hundreds = _____ ones

(b) 3000 = _____ hundreds = _____ ones

(c) 3300 = _____ hundreds = _____ ones

Problem 1.10 What number has

(a) 12 hundreds and 9 ones?

(b) 2 thousands, 21 tens, and 4 ones?

(c) 65 ones and 3 thousands?

1.3 Practice

Problem 1.11 Fill in the blanks.

(a) _____ $+\,300\,+$ _____ $+\,7 = 2397$

(b) $2\,+$ _____ $+\,800\,+\,9000 = 9832$

(c) _____ $+\,700\,+$ _____ $+\,20 = 9728$

(d) $8\,+\,80\,+$ _____ $+\,8000 = 8888$

(e) _____ $+\,80\,+\,800 = 884$

Problem 1.12 Fill in the blanks.

(a) $8 + 400 + 8000 + 30 =$ _____

(b) $3 + 0 + 7000 + 10 =$ _____

(c) $8 + 60 + 600 + 7000 =$ _____

Problem 1.13 Find the number that has

(a) 0 thousands, 3 hundreds, 3 tens, and 6 ones

(b) 2 thousands, 0 hundreds, 0 tens, and 6 ones

(c) 5 thousands, 0 hundreds, 5 tens, and 3 ones

Problem 1.14 What 3-digit numbers number have 1 ten and 1 more one than tens?

Problem 1.15 What number has 0 thousands, 8 hundreds, 1 more ten than hundreds, and no ones?

Problem 1.16 If the ones digit of a number is 4, the tens digit is twice the ones digit, and the hundreds digit is half of the ones digit, what is the number?

Problem 1.17 What is the difference between the largest and the smallest 5-digit numbers, each greater than 40000, that can be formed using only the digits 2, 3, 4, 5 and 6 if no digit is used more than once?

Problem 1.18 What 4-digit number has 2 tens, 6 hundreds, 1 more than the hundreds digit on the thousands digit, and 2 less ones than hundreds?

Problem 1.19 Find a 3-digit number such that the tens digit is a multiple of the hundreds digit and the ones digit is a multiple of the tens digit.

Problem 1.20 A number has 4 digits. Its ones digit is the second highest one digit number. The tens digit is half the ones digit. The hundreds digit is one more than half of the tens digit. The thousands digit is the difference between the tens digit and the hundreds digit. What is the number?

Problem 1.21 There is a 3-digit number. Its ones digit is 5 times its hundreds digit. Its tens digit is the sum of its ones digit and its hundreds digit. What is the 3-digit number?

Problem 1.22 There is a 2-digit number. The sum of the two digits is 7. If you switch the two digits, the new number is 27 more than the original 2-digit number. What is the original 2-digit number?

Problem 1.23 What is the difference between the greatest 5-digit number and the smallest 3-digit number?

Problem 1.24 Write the greatest 4-digit number using all different digits with a 6 in the tens place.

Problem 1.25 We want to write some 7 digit numbers. Let's pretend we can only use the digits 5, 1, 6, 2, 0, 9, and 7, and we can only use each digit once.

  (a) What is the largest number we can write?

  (b) The smallest number we can write?

(c) The largest odd number?

(d) The smallest even number?

Problem 1.26 Use the digits 0, 2, 3, 5, 7 to make two 5-digit numbers such that the difference of the two numbers is as large as possible. What is the difference?

Problem 1.27 Uncle Jim got lost while we were driving back to California from Montréal. He saw a sign that said how many miles we were from Los Angeles. Since he was driving fast, he couldn't quite see the number, but he knew it had 4 digits. I saw the exact number, but I wanted to have some fun so instead of telling him the number right away I gave him some clues:

- The number has the digit 1 somewhere.
- The digit in the hundreds place is three times the digit in the thousands place.
- The digit in the ones place is 4 times the digit in the tens place.
- The thousands digit is 2.

How far away are we from Los Angeles?

Problem 1.28 George's family lives in a house with a 4-digit street number. The difference between the first digit and the last digit is 8. The 2nd digit is twice the first digit, and the 3rd digit is twice the 2nd digit. What is the street number of George's house?

Problem 1.29 There were as many kids in the first car of the roller coaster as the largest possible sum of two different 1-digit numbers. How many kids were in the first car of the roller coaster?

Problem 1.30 The number of pieces of candy I ate last year is the same as the largest even number smaller than 1500. How many pieces of candy did I eat?

Problem 1.31 I have a toy car collection that has as many cars as the smallest odd number bigger than 3900. How many cars do I have in my collection?

Problem 1.32 Ron wants to go to the arcade so he will need as many quarters as possible. His dad gave him two $5 bills, and his mom gave him three $1 bills and one $10 bill. How many quarters can he get with all the money he has?

Problem 1.33 Travis has been saving all the change he gets whenever he uses a $1 bill. After one month he has gathered 150 pennies, 53 nickels, and 7 quarters. Does he have enough money to buy a popsicle that costs $5.99?

Problem 1.34 Can you find 5 consecutive 2-digit numbers such that their sum has 6 tens and no hundreds?

Problem 1.35 Find three consecutive numbers with 2 hundreds whose sum is even.

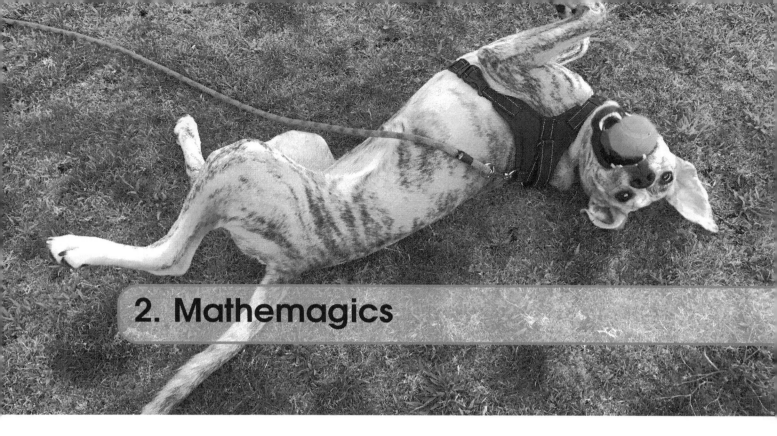

2. Mathemagics

Ms. Patsy needs to add the scores that 10 of her students received on a test. Her students' scores were:

$$92, 86, 91, 88, 87, 90, 89, 93, 92 \text{ and } 88.$$

She is in a hurry, can you help her out?

Have you ever wondered how some people can do arithmetic calculations within seconds? We want to show you some useful tricks for arithmetic in this chapter. With some practice many of these strategies can be done in your head. Pick out your favorites and try to come up with your own tricks and shortcuts!

The concepts introduced in this chapter directly correspond to Common Core Math Standards as shown in the following table.

3rd Grade	3.OA.1, 3.OA.2, 3.OA.5, 3.NBT.2
4th Grade	4.NBT.1, 4.NBT.2, 4.NBT.5

In addition to the standards above, problems and concepts in this section will help strengthen understanding of the following domains.

3rd Grade	3.OA, 3.NBT
4th Grade	4.OA, 4.NBT
5th Grade	5.OA, 5.NBT

2.1 Example Questions

Example 2.1

Ms. Patsy needs to add the scores that 10 of her students received on a test. Her students' scores were:

$$92, 86, 91, 88, 87, 90, 89, 93, 92 \text{ and } 88.$$

She is in a hurry, can you help her out?

Solution

We want to find the sum

$$92 + 86 + 91 + 88 + 87 + 90 + 89 + 93 + 92 + 88.$$

As Ms. Patsy is in a hurry, we should do this as quick as possible. Apparently all the students did great: all of the scores are close to 90 points. Let's use that to our advantage: instead of adding them right away, we will find out *how far they are from* 90 and use that to to add them up quicker. We can rewrite the sum using numbers that would give us the original numbers if we add them to, or subtract them from, 90. In this case, our quick sum would look like:

$$2 - 4 + 1 - 2 - 3 + 0 - 1 + 3 + 2 - 2,$$

and we have some numbers that we can cancel right away:

$$\cancel{2} - 4 + \cancel{1} - \cancel{2} - \cancel{3} + 0 - \cancel{1} + \cancel{3} + \cancel{2} - \cancel{2} = -4.$$

Since we originally had 10 numbers, we will need to subtract this 4 from $10 \times 90 = 900$. This gives us that the total sum is $900 - 4 = 896$.

Example 2.2

Katrine and Mindy work at a craft store. A customer is buying some long ribbons of different lengths, and they need to know how much he is buying in total. The lengths of each of the ribbons are $54\,cm$, $67\,cm$, $33\,cm$, $84\,cm$, $46\,cm$, $64\,cm$ and $16\,cm$. How many cm of ribbon did the customer buy in total?

Solution

We need to add up the lengths of all the pieces of ribbon, so we want to find

$$54 + 67 + 33 + 84 + 46 + 64 + 16.$$

Notice we have some pairs numbers that add up to 100. We can spot them easily because their units digits add up to 10 and their tens digits add up to 9.

$$\underline{54} + \underline{67} + \underline{33} + \underline{84} + \underline{46} + 64 + \underline{16}$$

This means we can add the numbers a lot quicker if we replace each of the pairs $(54 + 46, 67 + 33, 84 + 16)$ with a 100:

$$100 + 100 + 100 + 64 = 364$$

The customer bought $364\,cm$ of ribbon.

Remark

> There may be some times that you can spot pairs of numbers that add up to a common number. Always use that to your advantage!

Example 2.3

Some times we will be able to save time if we find that some of the digits of the number we are subtracting match those of another number.
   (a) $2947 - 347$
   (b) $8272 + 1728 - 172 - 728$

Solution to Part (a)

Since both numbers end up in 47, we just need to worry about the first digits

$$2947 - 347 = 2900 - 300 = 2600$$

Solution to Part (b)

Again, some of the numbers end with the same digits. Grouping these numbers we can get rid of the last digits right away:

$$\underbrace{8272 - 172}_{8100} + \underbrace{1728 - 728}_{1000} = 9100$$

Example 2.4

Let's multiply numbers by 11!
   (a) What is 253×11?
   (b) What is 3594×11?
   (c) What is 45729×11?

Solution to Part (a)

When we multiply a number by 11, we can speed up the process by adding up neighbor digits. It will be even easier if we start from the right. The last digit, 3, has no neighbor to the right, so we can pretend it had a 0 neighbor and add $3 + 0 = 3$. Then we have 5 and 3 are neighbors, so we add them together to get $5 + 3 = 8$. The next pair of neighbors is 2 and 5, $2 + 5 = 7$. And lastly, the 2 has no neighbor to the left, so we can pretend it has a 0 neighbor, so $0 + 7 = 7$. The process is summarized in the following diagram:

$$
\begin{array}{ccccccc}
2 & & 5 & & 3 & & \\
\swarrow & \searrow & \swarrow & \searrow & \swarrow & \searrow & \\
2 & 2+5 & & 5+3 & & 3 & \\
2 & 7 & & 8 & & 3 &
\end{array}
$$

These sums allow us to "read" out our answer, digit by digit, so

$$253 \times 11 = 2783.$$

Solution to Part (b)

Sometimes, when we add up neighbors we will get a number that is bigger than 9, so it has more than one digit. In those cases we will need to carry over the extra digit to the *left* neighbor sum. That is why it is easier if we start adding neighbors from the right!

$$
\begin{array}{ccccccc}
3 & & 5 & & 9 & & 4 \\
\swarrow & \searrow & \swarrow & \searrow & \swarrow & \searrow & \\
3 & 3+5 & 5+9 & 9+4 & & 4 & \\
3 & 8 & 14 & 13 & & 4 & \\
3 & 9 & 5 & 3 & & 4 &
\end{array}
$$

Solution to Part (c)

Sometimes we will need to carry more than once, but this isn't a probem!

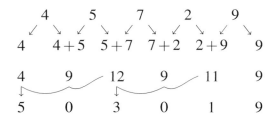

Example 2.5

"Teen" numbers are the 2-digit ones that have one ten, so all numbers between 10 and 19. There is a shortcut to multiply two of these numbers. Try the following:
 (a) 16×17
 (b) 19×13

Solution to Part (a)

Here are the steps we want to follow for this shortcut:

Step 1 Multiply the ones digits, keeping the resulting ones digit and making a note of how many tens, as we'll use them in Step 2.

Step 2 Add 10 to the ones digits of both numbers and add to this the number of tens you got from Step 1.

Step 3 Append the ones digit you got in Step 1 to the number you got in Step 2. This gives the answer.

Let's see it in action!

 1. The product of the ones digits is $6 \times 7 = 42$, which has 4 tens and 2 ones.
 2. $10 + 6 + 7 + 4 = 27$
 3. The answer is 272.

Solution to Part (b)

 1. $9 \times 3 = 27$
 2. $10 + 9 + 3 + 2 = 24$
 3. The answer is 247.

Example 2.6

Let's multiply some multi-digit numbers!
  (a) 97×48
  (b) 354×293
  (c) 287×94

Solution to Part (a)

An easy way to multiply numbers is setting up a multiplication table like this one:

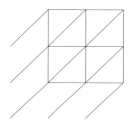

We are using a table with 2 rows and 2 columns because we are multiplying a 2-digit number by a 2-digit number. The diagonal lines will help give us the digits of our answer during the process.

We will start by writing each of the digits of our first number at the top of each column, and each of the digits of our second number at the right of each row.

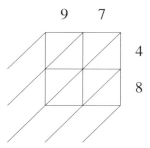

In each of the squares inside we will write the product of the corresponding digits. Here we have $9 \times 4 = 36$, $7 \times 4 = 28$, $9 \times 8 = 72$, and $7 \times 8 = 56$. We will write the *tens digit* in the upper triangle and the *ones digit* in the lower triangle.

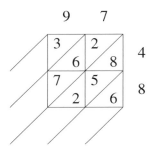

We will then proceed to add all the digits we have in each of the diagonals, starting with the bottom right. We want to end up with one single digit in each diagonal, so we will carry over the extra to the diagonal to the left. Here we get $6 = 6$, $2 + 5 + 8 = 15$, so carrying the one $1 + 7 + 6 + 2 = 16$, so carrying the one $1 + 3 = 4$.

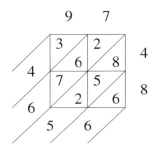

We will be able to read the product of our numbers by following the digits on each diagonal from left to right. In this case we get that $97 \times 48 = 4656$.

| Remark |

> This multiplication table presents in an organized manner all the single digit multiplications we would normally have to do using the traditional method. However, we do not have to worry about carrying over at the same time we multiply the numbers.

Solution to Part (b)

We can use the same kind of table, but we will need it to be bigger since we are multiplying 3-digit numbers:

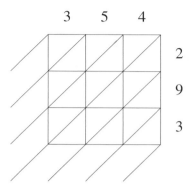

We need to be careful when we end up with 1 digit numbers when we multiply. Since 1-digit numbers have 0 tens, we will write a 0 in the tens' triangle whenever we get a 1-digit number after we multiply.

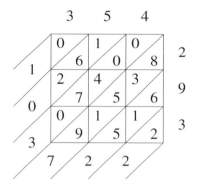

This means, $354 \times 293 = 103722$.

Solution to Part (c)

We can also use this kind of multiplication tables when we multiply numbers that do not have the same number of digits. This time we will use a table with 3 columns and 2 rows.

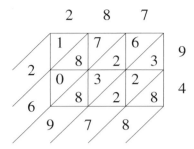

So we have that $287 \times 94 = 26978$.

Example 2.7

Calculate the following squares:
 (a) 15^2
 (b) 165^2
 (c) 2005^2
 (d) 13^2
 (e) 68^2
 (f) 104^2

Solution to Part (a)

When squaring a number that ends in 5, the number we get will *always* end in 25 because $5 \times 5 = 25$. We can figure the rest of the digits by multiplying the number we get by removing the 5 together with its successor (that is, the next counting number).

In the number 15 when we remove the 5, we are left with 1. So we need to multiply $1 \times 2 = 2$ and then attach 25 at the end of it, giving $15^2 = 225$:

Solution to Part (b)

The number 165 without the 5 at the end is just 16, and the successor of 16 is 17. So, we need to multiply $16 \times 17 = 272$ and attach a 25 at the end:

Solution to Part (c)

This strategy also works with big numbers. If we remove the 5 at the end of 2205 we are left with 220, so to find 2005^2 we would just need to multiply $200 \times 201 = 40200$ and attach a 25 at the end of it:

$$
\begin{array}{c}
2005 \\
200 \times 201 \\
4020025
\end{array}
$$

So, $2005^2 = 4020025$.

Solution to Part (d)

What about numbers that don't end in 5? If you follow these steps it will be easier (and faster!) to find the squares of some numbers (that is, if you haven't memorized the answer already).

Step 1 Start by finding the closest number to 13 that ends in 0, in this case that number is 10.

Step 2 To get 10 from 13 we would need to subtract 3, so now add 3 to 13 to get 16.

Step 3 Multiply the numbers you got from Step 1 and Step 2, that is $10 \times 16 = 160$.

Step 4 To the number you got in Step 3 add the square of the difference you used in Steps 2, so $3^2 = 9$, to get your answer $160 + 9 = 169$

Once you get comfortable with this method, the steps are easy to perform all at once:

$$13^2 = (13 - 3) \times (13 + 3) + 3^2 = 10 \times 16 + 9 = 169$$

so we see $13^2 = 169$.

Remark

This works because of a useful tool you will learn in a couple of years in your algebra class called *difference of squares*. The rule says that if you subtract the squares of two numbers, that is the same as if you multiply their sum and their difference. They usually write that rule like this:

$$a^2 - b^2 = (a + b)(a - b).$$

We are using a as the number we want to square, and b is the distance to the closest number to it that ends in 0.

Solution to Part (e)

This time the closest number to 68 that ends in 0 is 70. Since we add 2 to 68 to get 70, this time we will subtract 2 instead of adding in Step 2, to get $68 - 2 = 66$. So we need to multiply 70 and 66 to get 4620 and then add $2^2 = 4$. So our final answer is 4624.

$$68^2 = (68 + 2) \times (68 - 2) + 2^2 = 70 \times 66 + 4 = 4624$$

Solution to Part (f)

This strategy will work even if we have numbers with more than 2 digits. The closest number to 104 that ends in 0 is 100, which we get by subtracting $104 - 4$. We will then multiply 100 by $104 + 4 = 108$ and add $4^2 = 16$ to that to get our answer.

$$104^2 = (104 - 4) \times (104 + 4) + 4^2 = 100 \times 108 + 16 = 10816$$

Example 2.8

Now try multiplying these numbers! Note that they start with the same digits and end in digits that add up to 10.

    (a) 48×42

    (b) 39×31

    (c) 264×266

Solution to Part (a)

If you notice that two numbers that you are multiplying start with the same digits and end up with digits that add up to 10, you can use the following shortcut to find their product:

Step 1 Remove the last digit from your number and multiply the number you get by its successor, that is, the next counting number.

Step 2 Multiply the ones digits and append the number you get to the number you got in Step 1.

Now let's see it in action!

1. We need to multiply 4 and 5 to get 20.
2. The product of the ones digits is $8 \times 2 = 16$, so our answer is 2016.

Solution to Part (b)

We want to be careful when we get a one digit number in Step 2 (in this case our answer will end in 09, not just 9). This time we want to multiply 3 and 4 to get 12, and in the end append a 09, so our answer is 1209.

Solution to Part (c)

This will still work fine, no matter how many digits we have, as long as the ones digits add up to 10 and all the other digits in the number are the same. So we do first $26 \times 27 = 702$ and attach a $4 \times 6 = 24$ in the end to get 70224.

Example 2.9

What is the easiest way to multiply and divide these numbers?
  (a) $213 \times 24 \div 12$
  (b) $75 \times 32 \div 400$
  (c) $36 \times 62 \div 2 \times 7 \div 3$

Solution to Part (a)

If we were to do these operations in order from left to right we would work with bigger numbers:

$$213 \times 24 \div 12 = 5112 \div 12 = 426.$$

We want to try to make the process simpler by working with smaller numbers. Note that we have a 24 that is multiplying and a 12 that is dividing. Since $24 \div 12 = 2$, if we do that first we will end up with

$$213 \times 24 \div 12 = 213 \times 2 = 426.$$

Remark

Careful! If you have two $\div$ signs in a row you cannot do the second division first. For example, in

$$32 \div 16 \div 2 = 2 \div 2 = 1$$

we would get the WRONG number if we start dividing by 16 and 2:

$$32 \div 16 \div 2 \neq 32 \div 8 = 4 \qquad \longleftarrow \text{ Not the right answer}$$

Solution to Part (b)

Note that both 32 and 400 are even numbers (so, they are both divisible by 2) and one of them is multiplying while the other is dividing. We now use this observation to simplify our problem and use smaller numbers. We can even do this more than once!

$$\frac{75 \times \overset{16}{\cancel{32}}}{\underset{200}{\cancel{400}}} = \frac{75 \times \overset{8}{\cancel{16}}}{\underset{100}{\cancel{200}}} = \frac{75 \times \overset{4}{\cancel{8}}}{\underset{50}{\cancel{100}}} = \frac{75 \times \overset{2}{\cancel{4}}}{\underset{25}{\cancel{50}}} = \frac{\overset{3}{\cancel{75}} \times 2}{\cancel{25}} = 3 \times 2 = 6$$

Solution to Part (c)

This time we have more than one number dividing. Let's put the numbers in two groups: in one group all the numbers that are multiplying and on the other all the numbers that are dividing. For convenience we will write the division as one large horizontal line, that way we can easily spot if there is anything we can cancel.

$$\frac{36 \times 62 \times 7}{2 \times 3}.$$

We can divide 36 by 3 and get 12 instead, and we can divide 62 by 2 and get 31 instead. So we have

$$\frac{\overset{12}{\cancel{36}} \times 62 \times 7}{2 \times \cancel{3}} = \frac{12 \times \overset{31}{\cancel{62}} \times 7}{\cancel{2}} = 12 \times 31 \times 7 = 372 \times 7 = 2604$$

We still had to deal with some big numbers, however, the numbers we worked with in the end were a lot smaller than if we had performed the operations in the order they were originally:

$$36 \times 62 \div 2 \times 7 \div 3 = 2232 \div 2 \times 7 \div 3 = 1116 \times 7 \div 3 = 7812 \div 3 = 2604$$

Remark

> You may have noticed that writing the numbers like this looks like working with fractions. This is no coincidence. When we are multiplying and dividing numbers, we can always write a fraction where the numerator has the product of all the numbers that are multiplying and the denominator has the product of all of the numbers that are dividing.

Example 2.10

Let's take a look at a shortcut for multiplying and dividing by 5 and 25.
 (a) 323×5
 (b) 6248×25
 (c) $1605 \div 5$
 (d) $1825 \div 25$

Solution to Part (a)

We can simplify the process of multiplying a number by 5 if we first multiply by 10 and then divide by 2, since $5 = 10 \div 2$. So

$$
\begin{aligned}
323 \times 5 &= 323 \times 10 \div 2 \\
&= 3230 \div 2 \\
&= 1615
\end{aligned}
$$

Solution to Part (b)

When we multiply by 25 we can do something similar. This time we will multiply by 100 and divide by 4, as $25 = 100 \div 4$. When we divide by 4, it is sometimes easier to divide by 2 and then divide by 2 one more time

$$
\begin{aligned}
6248 \times 25 &= 6248 \times 100 \div 4 \\
&= 624800 \div 4 \\
&= 624800 \div 2 \div 2 \\
&= 312400 \div 2 \\
&= 156200
\end{aligned}
$$

Remark

Note that in the last example we could start off dividing by 4 instead of multiplying by 100, since 6248 is divisible by 4.

$$
\begin{aligned}
6248 \times 25 &= 6248 \div 4 \times 100 \\
&= 1562 \times 100 \\
&= 156200
\end{aligned}
$$

In the first example it would not work to first divide by 2 and then multiply by 10 because 323 is not an even number.

Solution to Part (c)

This time we are dividing by 5, but we can still do something similar. We can multiply by 2 and divide by 10:

$$1605 \div 5 = 1605 \times 2 \div 10 = 3210 \div 10 = 321$$

Solution to Part (d)

You guessed right! Instead of dividing by 25 we can multiply by 4 and divide by 100:

$$1825 \div 25 = 1825 \times 4 \div 100 = 7300 \div 100 = 73$$

2.2 Quick Response Questions

Problem 2.1 Write 5 pairs of 1-digit numbers such that the sum of each pair is 10.

Problem 2.2 Write 5 pairs of 1-digit numbers such that the sum of each pair is 9.

Problem 2.3 Use the pairs you found in the previous question to write 10 pairs of 2-digit numbers that add up to 99. Try to be creative and include numbers that start with different digits!

Problem 2.4 Write 10 pairs of 2-digit numbers, that don't end in 0, such that the sum of each pair is 100.

Problem 2.5 Write 10 pairs of 3-digit numbers, that don't end in 00, such that the sum of each pair is 1000.

Problem 2.6 Group the numbers 6, 9, 11, and 8 into pairs so that each pair has the same sum.

Problem 2.7 Find the product of the following "teens".

(a) 19×12

(b) 16×14

(c) 13×18

Problem 2.8 Multiply the following numbers by 11

(a) 43

(b) 63

(c) 33

Problem 2.9 Find the squares of the following numbers.

(a) 18^2

(b) 19^2

(c) 17^2

Problem 2.10 Let's square some numbers! Try to do it as fast as you can! (This does not mean that we want you to break your pencil because you are writing too fast, but remember we learned a technique to do these calculations quicker!)

(a) 65^2

(b) 95^2

(c) 75^2

2.3 Practice

Problem 2.11 All the following pairs of numbers have something in common. See if you can identify what that is and find their sums!

(a) $70 + 29$

(b) $462 + 537$

(c) $468 + 531$

(d) $7755 + 2244$

(e) $8252 + 1747$

Problem 2.12 Add up the following numbers. Try to identify pairs of numbers that add up to 9, 99, 999, 9999, 10, 100, 1000 or 10000 to make your job easier!

   (a) $326 + 674 + 54 + 45$ (Find a pair of numbers that adds up to 1000 and another pair that adds up to 99).

   (b) $5694 + 4306 + 598$ (Find a pair of numbers that adds up to 10000).

   (c) $6845 + 3154 + 4987 + 5013$ (Find a pair that adds up to 10000 and another pair that adds up to 9999).

Problem 2.13 Times 11 time! Multiply the following numbers by 11. Remember to add up neighbors to save time!

   (a) 743×11

   (b) 6172×11

(c) 3241×11

Problem 2.14 Some of the numbers in the following questions are close to numbers that are easier to work with. Use that to your advantage when calculating!

(a) $1456 - 299$

(b) $19998 + 3 + 1999 + 998 + 3 + 999$

(c) $49996 + 39993 + 29992 + 19991 + 998$

Problem 2.15 Before you add the numbers, find pairs of numbers that add up to something that ends with a 0. That should make the calculation a lot easier!

(a) $191 + 809 + 259 + 2329 + 1741$

(b) $137 + 356 + 863 + 644$

(c) $829 + 571 - 692 - 308$

Problem 2.16 Take a look at the last two digits of the numbers in the following. Find something in them that may help you do this computations quicker.

(a) $673 + 528 - 373 + 472$

(b) $2273 - 655 - 345 - 273$

(c) $2948 + 4355 - 648 - 155 + 74$

Problem 2.17 Use the strategy for squaring numbers that end up in 5 to find out the squares of this numbers. Remember that the square of a number that ends in 5 always ends in 25! Hint: We also learned a strategy for multiplying "teens" that may help!

(a) 105^2

(b) 115^2

(c) 125^2

(d) 135^2

(e) 145^2

(f) 155^2

(g) 165^2

(h) 175^2

(i) 185^2

Problem 2.18 Fill in the blanks

(a) $78 = 25 \times \underline{\hspace{1cm}} + \underline{\hspace{1cm}}$

(b) $277 + 177 = 25 \times \underline{\hspace{1cm}} + 4 = \underline{\hspace{1cm}}$

(c) $273 + 152 + 25 = 25 \times \underline{\hspace{1cm}} = 50 \times \underline{\hspace{1cm}} = \underline{\hspace{1cm}}$

Problem 2.19 In each of these sums, the summands are each close to a number. Identify that number to fill in the blanks and find the sum.

(a) $12 + 13 + 9 + 8 + 13 + 9 = 10 \times \underline{\hspace{1cm}} + \underline{\hspace{1cm}} = \underline{\hspace{1cm}}$

(b) $98 + 105 + 94 + 101 + 99 = 100 \times \underline{\hspace{1cm}} - \underline{\hspace{1cm}} = \underline{\hspace{1cm}}$

(c) $997 + 1004 + 1003 + 996 + 999 + 1005 = 1000 \times \underline{\hspace{1cm}} + \underline{\hspace{1cm}} = \underline{\hspace{1cm}}$

Problem 2.20 Identify a number that is close to all the numbers in the sum. Use that number to help find the sum.

(a) $47 + 56 + 43 + 60 + 43 + 52 + 40 + 48 + 45 + 58 + 41 + 55$

(b) $504 + 505 + 510 + 499 + 504 + 492 + 501 + 503 + 490 + 492 + 501 + 501 + 499$

(c) $1505 + 1509 + 1497 + 1499 + 1506 + 1496$

Problem 2.21 In the following sums, pairs of numbers add up to a common number. Use that to your advantage.

(a) $57 + 13 + 9 + 29 + 41$

(b) $70 + 50 + 33 + 12 + 69 + 51 + 108$

(c) $99 + 13 + 86 + 24 + 16 + 75$

Problem 2.22 Find pairs of numbers that add up to the same number. Then use these numbers to add and subtract quicker. Be careful with the minus signs!

(a) $29 + 46 - 28 - 47 + 38 + 37 + 44 + 31$

(b) $15 + 47 + 42 + 20 - 30 - 45 - 32 - 17 + 46 + 16$

(c) $21 + 17 + 24 + 28 + 23 + 22 - 16 + 18 + 16 + 29 + 27 - 29$

Problem 2.23 Dulce is going to order an enormous amount of croissants for some upcoming breakfast meetings. Since she is ordering so many croissants, the baker offered her a deal that for every box of croissants she buys she gets 1 extra croissant for free. If croissants normally come in boxes of 10, how many croissant (including the free croissant per box) wlil Dulce get if she orders

(a) 140 boxes of croissants?

(b) 3205 boxes of croissants?

Problem 2.24 Coach Robertson is setting up a huge soccer tournament and is inviting teams from around the whole country. He needs to set up hotel accommodations for all the teams attending. If every team will bring a total of 22 people and each room fits 2 people, how many rooms does he need to book if 365 teams come to the tournament?

Problem 2.25 Donna needs to buy some office supplies for her company. The supplier has a deal that for every 5 boxes she buys, she will get 1 extra box of the same supplies for free.

(a) How many pens did she get for free if she paid for 140 boxes of 10 pens?

(b) How many envelopes did she get for free if she paid for 1805 boxes of 100 envelopes?

Problem 2.26 Ms. Pelegrino is grading some tests. There are 20 questions on each of the 5 pages of the test. Apparently her students did great on the test. So far she has the number of correct answers on each page of the test. Help her find out the total scores for the following two tests, based on the scores on each page.

(a) Correct answers per page: 20, 18, 19, 18 and 20.

(b) Correct answers per page: 17, 20, 18, 15 and 17.

Problem 2.27 Sam is using colored square tiles to cover some canvases for his art class. The tiles are 1 inch long per side. How many tiles will he need for each of the canvases?

(a) A square canvas that is 17 inches long per side.

(b) A square canvas that is 35 inches long per side.

(c) A rectangular canvas that is 12 inches long and 22 inches wide.

Problem 2.28 Compute the following. Before multiplying and dividing, make sure to group the numbers to simplify the calculation.

(a) $50 \times 25 \div 2$

(b) $350 \times 22 \div 7 \div 2$

(c) $715 \div 44 \times 4$

Problem 2.29 Find the product of the following pairs of numbers. Notice that they start with the same digits and their last digits add up to 10.

(a) 67×63

(b) 24×26

(c) 83×87

Problem 2.30 Find the product of the following pairs of numbers. Notice that they start with the same two digits and their last digits add up to 10.

(a) 497×493

(b) 311×319

(c) 783×787

Problem 2.31 Group together the numbers that are multiplying and the numbers that are dividing. Write the expression using the horizontal line as division. Try to identify ways to simplify the numbers before you proceed to multiply and divide.

(a) $63 \times 12 \div 7 \div 9 \times 4$

$$\frac{\quad\times\quad\quad\times\quad}{\times} =$$

(b) $25 \times 3 \div 5 \div 2 \times 64$

$$\frac{ \times \times }{\times} =$$

(c) $4 \div 3 \times 36 \div 10 \times 50$

$$\frac{ \times \times }{\times} =$$

Problem 2.32 With the aid of the given multiplication tables, find the product of the numbers.

(a) 67×34

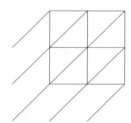

(b) 93×82

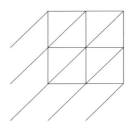

(c) 74×29

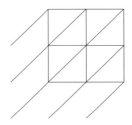

Problem 2.33 With the aid of the given multiplication tables, find the product of the numbers.

(a) 124×394

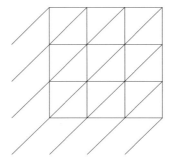

(b) 567×291

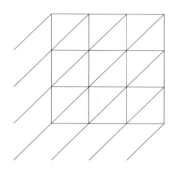

(c) 696×313

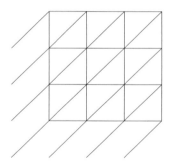

Problem 2.34 Use the rectangular multiplication tables to find the product of the given numbers.

(a) 67×2369

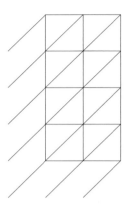

(b) 6457×34

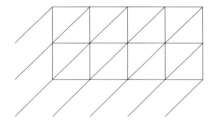

(c) 3366×821

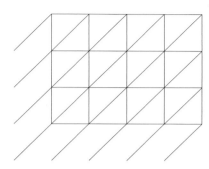

Problem 2.35 Use the rectangular multiplication tables to find the product of the given numbers.

(a) 13×481

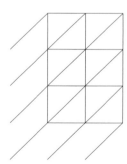

(b) 761×34

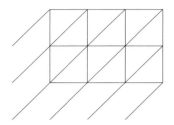

(c) 49×273

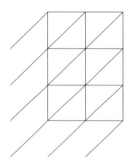

3. Big and Small, Included

You are invited to visit Old Mr. Farmer Guy at his farm. When you arrive, there is a party going on with all his roosters and hens. They are running all over the place so it is difficult to count how many there are. Each of the chickens is wearing a party hat and Mr. Farmer Guy tells you that he was missing 45 hats from his secret "party hat" stash. He's pretty sure that he has 4 times as many hens as roosters. Can you figure out how many roosters and how many hens he has?

Looking at a checkerboard, we see there are 64 squares, half colored white and the other half colored black. We know that 32 are white and 32 are black. How do we get from 64 to 32? What if for every white square there were three black squares? In this chapter we practice using ratios and grouping to help solve a wide variety of questions!

The concepts introduced in this chapter directly correspond to Common Core Math Standards as shown in the following table.

3rd Grade	3.OA.3, 3.OA.8, 3.MD.3
4th Grade	4.OA.1, 4.OA.2, 4.OA.3

In addition to the standards above, problems and concepts in this section will help strengthen understanding of the following domains.

3rd Grade	3.OA, 3.NBT, 3.MD
4th Grade	4.OA, 4.NBT, 4.MD
5th Grade	5.OA, 5.NBT, 5.MD

3.1 Example Questions

Example 3.1

Let's help out Old Mr. Farmer Guy figure out how many roosters and how many hens he has. Remember he said he was pretty sure he had 4 times as many hens as roosters and he is missing 45 party hats.

Solution

Let's put the chicken in small groups. We know that for each rooster we have 4 hens, so we can try to find out how many groups of 5 chicken (1 rooster and 4 hens) we can make. Since there are 45 party hats missing, that means we have 45 chicken in total, so we can make exactly $45 \div 5 = 9$ groups of 5 chickens. As we have 1 rooster and 4 hens in each small group, there are 9 roosters and $9 \times 4 = 36$ hens.

Example 3.2

In a school, there are 165 students in the third and fourth grades altogether. If there were 6 more third graders, the third grade would have twice as many students as the fourth grade. How many students are there in each of the two grades?

Solution

If we had 6 more third graders, we would have $165 + 6 = 171$ students in total. In that case, for each fourth grader we would have 2 third graders and we would be able to make $171 \div (1 + 2) = 57$ groups of 3 students. That would mean that we had 57 fourth graders and $57 \times 2 = 114$ third graders. Now, since we were assuming we had 6 extra students in the third grade, we actually have $114 - 6 = 108$ third graders and 57 fourth graders.

Example 3.3

Coach just brought a *huge* bag with 75 balls for gym class. He told us that in the bag there are twice as many basketballs as soccer balls, and that there are 3 more volleyballs than soccer balls. He won't let us play with them until we figure out exactly how many of each he brought. Can you help us out?

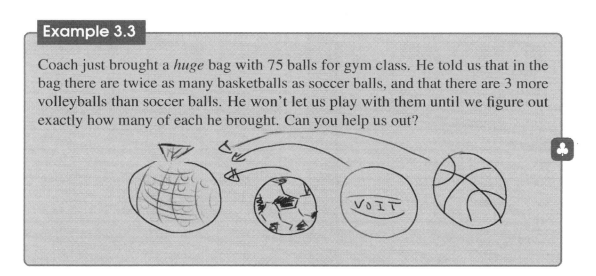

Solution

If we got rid of the 3 *extra* volleyballs, we would have $75 - 3 = 72$ balls in total and we would have the *same* number of soccer balls and volleyballs. Since we know that we have 2 basketballs for each soccer ball, we can make groups of 4 balls: 1 soccer ball, 1 volleyball and 2 basketballs. So, we have exactly $72 \div 4 = 18$ groups of 4 balls. That means we have 18 soccer balls, 18 volleyballs and $18 \times 2 = 36$ basketballs. Now we just need to put back the 3 volleyballs we pretended we didn't have, and so, we actually have $18 + 3 = 21$ volleyballs.

Example 3.4

Alice, Beth, and Cynthia have $50 in total. Alice has twice as much money as Beth, and Beth has 3 times as much as Cynthia. How much money does each ♣ have?

Solution

Alice has more money than Beth, and Beth has more money than Cynthia, so Cynthia is the one that has the least amount of money and Alice is the one that has the most money. We can pretend that Alice, Beth and Cynthia have all their money in $1 bills, so they have exactly 50 bills. For each bill that Cynthia has, Beth will have two, and for each bill that Beth has, Alice will have three. So, for each bill that Cynthia has, Alice will have $2 \times 3 = 6$ bills. Take some empty envelopes an put in each of them 1 bill for Cynthia, 3 bills for Beth, and 6 bills for Alice. Each envelope then has $1 + 3 + 6 = 10$ bills. Since we have 50 bills in total, we will have $50 \div 10 = 5$ envelopes with $10 each. This way, Cynthia has 5 dollars, Beth has $3 \times 5 = 15$ dollars and Alice has $5 \times 6 = 30$ dollars.

Example 3.5

I just found a box with Christmas ornaments in the attic. It has a huge label on the front that says it contains 200 Christmas ornaments of three different colors (red, blue, and white). I'm so clumsy that I broke some of them by accident. It seems I broke 14 white ornaments and now I have the same number of white ornaments and blue ornaments. Mom says that before I broke them we had 4 more red ornaments than 3 times the number of white ornaments. How many red ornaments were there in the box before I found it? ♣

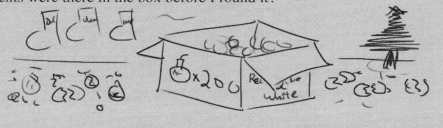

Solution

Let's rewind time and pretend no ornaments are broken. Also, let's put aside for now 4 of the red ornaments so we would have $200 - 4 = 196$ ornaments in total. This way the number of red ornaments is exactly the same as 3 times the number of white ornaments. It seems that there are 14 more white ornaments than blue ornaments. Let's pretend we have 14 more blue ornaments, so we have $196 + 14 = 210$ ornaments in total and the same amount of blue and white. This way, we can make groups of $3 + 1 + 1 = 5$ ornaments (3 red, 1 white and 1 blue). We can make $210 \div 5 = 42$ groups of 5 ornaments, so we have 42 white, 42 blue and $42 \times 3 = 126$ red. Since we assumed we had 14 more blue ornaments and we set aside 4 red ornaments, we actually have 42 white, $42 - 14 = 28$ blue and $126 + 4 = 130$ red ornaments.

Example 3.6

David has 3 times as much money as Chris. If David spends \$240, and Chris spends \$40 dollars, they will have the same amount of money. How much money do each of them have originally?

Solution

The difference between the amount of money David and Chris have is $240 - 40 = 200$ dollars. We can do something similar as in the previous problems to figure out how much money each of them have. We will take empty envelopes and put in them some \$1 bills. This time we know that for each bills Chris has, David will have 3. The one thing that will be different in this problem as in the previous ones, is that this time we know the *difference* of the amounts of money they have, so in each envelope we need to put 3 bills for David and *remove* 1 bill for Chris, so we will have $3 - 1 = 2$ bills in each envelope. We can make $200 \div 2 = 100$ envelopes like this, so Chris has 100 dollars and David has $100 \times 3 = 300$ dollars.

Example 3.7

Allison is 14 years old, and her dad is 50 years old. How many years ago was Allison's dad's age 5 times Allison's age?

Solution

The difference of Allison's age and her dad's age is $50 - 14 = 36$ years. Note that it is the same difference every year. Suppose we are now in the year when her dad's age is 5 times her age. We know that the difference of their ages is 36, and the ratio of their ages is 5, so Allison's age that year was $36 \div (5 - 1) = 9$. Since she is now 14 years old, that happened $14 - 9 = 5$ years ago.

Example 3.8

The school bought some basketballs, volleyballs, and soccer balls. There are 40 more soccer balls than volleyballs, and 8 fewer basketballs than volleyballs. Also, the number of soccer balls is 4 times the number of basketballs. How many balls of each kind are there?

Solution

We know the ratio between the number of basketballs and soccer balls, so we will try to find those two numbers first, and use them to find the third number. Since we have 40 more soccer balls *than volleyballs*, and 8 fewer basketballs *than volleyballs* the *difference* between the number of soccer balls and the number of basketballs is $40 + 8 = 48$. Knowing that for each basketball we have 4 soccer balls, we can find the number of basketballs if we divide by the ratio difference, $4 - 1 = 3$. So, we have $48 \div 3 = 16$ basketballs, $16 \times 4 = 64$ soccer balls, and $16 + 8 = 24$ volleyballs.

Example 3.9

You have $89 and your friend has $55. How much money should you give your friend so that he has exactly two times as much money as you?

Solution

You and your friend have $89 + 55 = 144$ dollars in total. You want your friend to have 2 dollars for every 1 dollar you have. This means you should have $144 \div (2 + 1) = 48$

dollars and your friend should have $38 \times 2 = 96$ dollars. In order to have 48 dollars, you should give your friend $89 - 48 = 41$ dollars.

Example 3.10

Tom and Jerry went fishing together and each caught some fish. If Tom gave one fish to Jerry, then they would have the same number of fish. If Jerry gave one fish to Tom, Tom would have 5 times as many fish as Jerry. How many fish did each of them catch?

Solution

We know the difference of the number of fish they have is 2, since they would have the same amount of fish if Tom gives one fish to Jerry. Let's pretend that Jerry gives 1 fish to Tom, this way Jerry has $2 + 2 = 4$ less fish, and Tom has exactly 5 times the number of fish that Jerry has. For every 1 fish that Jerry has Tom will have 5 fish, that is, 4 more fish than Jerry. We know that the difference in their number of fish is already 4, so Tom has 5 fish and Jerry has 1 fish. Remember we pretended that Jerry gave 1 fish to Tom, so Jerry actually has $1 + 1 = 2$ fish and Tom has $5 - 1 = 4$ fish.

3.2 Quick Response Questions

Problem 3.1 The sum of the ages of Aiden and Brian is 12. Aiden's age is twice Brian's age. What are their current ages?

Problem 3.2 Justin got 1 out of every 7 questions wrong on a 105 point test. How many questions did Justin get correct?

Problem 3.3 Adam gets \$1.75 for lunch every day. He bought a juice box for 25¢, and a burger for 5 times as much as the juice box. Does he have enough money left for an ice cream that costs 27¢?

Problem 3.4 Tom and Jerry went fishing. They caught 60 fish altogether. Tom caught 3 times as many fish as Jerry. How many fish did each catch?

Problem 3.5 In Old McDonald's corn field, 2 out of 3 scarecrows dance the Macarena. If there are 120 scarecrows in the field, how many don't do this dance?

Problem 3.6 David's dad bought a digital camera from an online store. He paid the price of the camera plus shipping and handling, for a total of $270. The price of the camera is 8 times the cost of shipping and handling. What is the price of the digital camera?

Problem 3.7 A country sent 108 athletes to the Olympic Games, among which the number of male athletes is twice the number of female athletes. How many male and female athletes are there?

Problem 3.8 Aiden and Brandon collected 69 rare coins altogether. Aiden collected 2 times more coins than Brandon. How many coins did each of them collect?

Problem 3.9 Suppose the height of an elf is 3 times the height of a hobbit. Also suppose that an elf is 60 inches taller than a hobbit. What is the height of each?

Problem 3.10 A farm has some ducks and geese. There are 8 more ducks than geese. The number of ducks is 3 times the number of geese. How many ducks and how many geese are there in the farm?

3.3 Practice

Problem 3.11 Your best friend is having a birthday party. Since his twin sisters' birthday is so close, their parents decided to have one party for the three of them at the same time. They ask you for help placing the candles on each of their cakes, but you can't remember how old each of them are becoming this year. They give you a total of 20 candles, and you also know that this year the age of your friend will be 3 times the age his twin sisters. How many candles should you place in each cake?

Problem 3.12 You have two big buckets of water: one is red and the other one is blue. The red bucket contains 14 liters of water, and the blue bucket contains 18 liters of water. If you want bucket the red bucket to contain 3 times as much water as the blue bucket, how much water do you need to pour from the blue bucket into the red bucket?

Problem 3.13 Monkey George had a basket of bananas. The total weight including the bananas and the basket is 32 pounds. Monkey George ate half of the bananas, and then the total weight including the bananas and the basket was 17 pounds. How many pounds of bananas were there originally? How heavy is the basket?

Problem 3.14 The sum of the ages of Adam and Bob is 14. Two years ago Adam's age was twice Bob's current age. What are their current ages?

Problem 3.15 Sinbad, Popeye, and Captain Hook went on the sea to hunt for treasure. They found totally 1645 gold coins. Popeye found twice as many as Sinbad, and Captain Hook found twice as many as Popeye. How many gold coins did each of them find?

Problem 3.16 The sum of the heights of Mr. Giant and Mr. Super Giant is 45 feet. If Mr. Super Giant is 5 feet taller than 4 times the height of Mr. Giant, what is the height of each of them?

Problem 3.17 There are 48 students in a class. If 3 more boys joined the class, the number of boys would be twice the number of girls. What is the current number of boys in the class?

Problem 3.18 Jadean and Lucy have $32 in total. Before Jadean bought a $3 pen, he had 4 times as much money as Lucy did. How much money did each of them have originally?

Problem 3.19 You have $20, and your friend has $25. How much money should your friend give you so that you will have twice as much money as your friend?

Problem 3.20 A pen and a pencil together cost $2.10. If the price of a pen is 6 times the price of a pencil, what is the price of each?

Problem 3.21 Sam has $30 more than 3 times the amount of money Tom has, and Paul has $15 less than Tom. Altogether they have $240. How much money does Sam have?

Problem 3.22 The sum of the ages of Lisa and Suzi is 24. Four years ago Lisa's age was three times Suzi's age. What are their current ages?

Problem 3.23 Adam has 55 songs in his iPod, including pop songs and classical music. The number of pop songs is 4 times as many as the number of pieces of classical music. How many pop songs and how many pieces of classical music does he have?

Problem 3.24 Bob's dad was 25 years old when Bob was born. The age of Bob's dad last year was 3 times Bob's current age. How old are Bob and his dad this year?

Problem 3.25 Old McDonald and Old Wendy went to the market to sell apples. They originally had the same number of pounds of apples. Old McDonald sold 11 pounds of apples, and Old Wendy sold 29 pounds. Now Old McDonalds has 3 times as many pounds of apples as Old Wendy. How many pounds of apples do each of them have now?

Problem 3.26 A farm produced $1,600$ pounds of fruit, including bananas and oranges. They produced 100 more pounds of bananas than 3 times the number of pounds of oranges they produced. How many pounds of bananas and how many pounds of oranges did they produce?

Problem 3.27 There are boys and girls in the classroom. Had there been 10 fewer boys, the numbers of boys and girls would have been equal. Had there been 10 fewer girls, the number of boys would have been twice the number of girls. How many boys, and how many girls are there?

Problem 3.28 Gold Toothed Brendan and Peg Legged Samuel went to hunt for treasures. Gold Toothed Brendan found 29 more diamonds than Peg Legged Samuel. Also, Gold Toothed Brendan found 1 more diamond than 3 times the number of diamonds that Peg Legged Samuel found. How many diamonds did each of them find?

Problem 3.29 Clark Kent and Neo are comparing their flying speeds. The speed of Clark Kent is 540 mph faster than that of Neo. The speed of Clark Kent is 90 mph less than 4 times the speed of Neo. What is the flying speed of each of them?

Problem 3.30 Bugs and Tweety were bored so they decided to throw pebbles into the lake. Originally they had the same number of pebbles. Bugs threw 7 pebbles and Tweety threw 19, then Bug had 3 times as many remaining pebbles as Tweety. How many pebbles do each of them have now?

Problem 3.31 Mickey, Donald and Goofy went for a bicycle ride. If Mickey rode 6 miles more than Donald, Goofy rode 22 miles more than Donald, and Goofy rode twice as far as Mickey, how far did each of them ride?

Problem 3.32 A baseball game is being played. Some people are watching at the stadium and some are watching on TV at home. The number of people who watch on TV is 480 more than the number of people in the stadium. If 50 people at the stadium went home and watched the game on TV, the number of people who watch on TV would be 5 times the number of the number of people in the stadium. How many people are watching the game in total?

Problem 3.33 Harry and Hermione are making magic potions. If Harry added 8 ounces of potion into his cauldron, he would have the same amount of potion as Hermione. If Harry poured 3 ounces of potion from his cauldron into Hermione's, then Hermione would have 3 times as much potion as Harry. How much potion do each of them have originally?

Problem 3.34 A room is rectangular in shape, and its length is 4 times its width. Given that the length is 36 feet longer than the width, find the area of the room in square feet.

Problem 3.35 You have $56, and your friend has $40. How much money should you give to your friend so that you two have the *same* amount of money?

4. Come Together, Leave Apart

Mary, Rose and Catherine are working on a school project and have some big pieces of ribbon of different sizes. If they glue together their pieces of ribbon they get a ribbon that is 840 centimeters long. Their teacher measured each of their ribbons and told them that Rose's ribbon is 130 centimeters longer than Mary's, and Catherine's is 220 centimeters longer than Rose's. Can you figure out how long is each ribbon?

Sometimes we know the sum of two numbers *and* their difference. We can use this information to figure out the numbers. In this chapter we'll learn some methods to quickly solve these types of problems, and see how to combine them with what you learned in the previous chapter. In many problems the sums and differences are not given right away, but with a little bit of investigation you can figure them out!

The concepts introduced in this chapter directly correspond to Common Core Math Standards as shown in the following table.

3rd Grade	3.OA.3, 3.OA.8, 3.MD.3
4th Grade	4.OA.1, 4.OA.2, 4.OA.3

In addition to the standards above, problems and concepts in this section will help strengthen understanding of the following domains.

3rd Grade	3.OA, 3.NBT, 3.MD
4th Grade	4.OA, 4.NBT, 4.MD
5th Grade	5.OA, 5.NBT, 5.MD

4.1 Example Questions

Example 4.1

You found a bag with some balls of different colors in the garage. It seems there are only black and white balls in there. Your mom says there are 10 balls in total and that she remembers having 2 more black balls than white balls in there. How many balls of each color are there in the bag?

Solution

We have 10 balls in total that are either black or white.

⑦ ⑦ ⑦ ⑦ ⑦ ⑦ ⑦ ⑦ ⑦ ⑦

We know that there are 2 more black balls than white balls, so let's color two of those balls black.

We have still 8 balls that could be either white or black, but we should have the same amount of each. Let's split them 4 and 4!

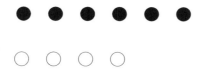

So, we have 6 black balls and 4 white balls.

Remark

This was a simple problem that used small numbers, so that is why we could draw the balls and color them to find out how many of each we had. But... what if we had bigger numbers? We don't want to be drawing 1000 balls and coloring them one by one!

Take a look at the next problem. It uses bigger numbers, and we can still do something *similar* to what we just did.

Example 4.2

A school bought some basketballs and volleyballs, 50 balls in total. There are 10 more basketballs than volleyballs. How many balls of each kind are there?

Solution

This time we are working with bigger numbers, and we don't want to draw 50 balls to figure out how many of each there are. Instead we will draw some long bars that will represent the number of balls of each kind that we have. The bar for the number of basketballs will be longer because we have 10 more basketballs than volleyballs:

If we had 10 more Volleyballs, we would have the same amount of balls of each and we would have $50 + 10 = 60$ balls in total.

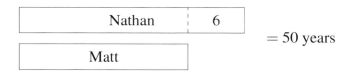

So, if we count the Basketballs alone, we should have $60 \div 2 = 30$ of them! Which leaves us with $30 - 10 = 20$ Volleyballs.

Example 4.3

When Matt was 17 years old, Nathan was 23. This year the sum of their ages is 50. What are their ages this year?

Solution

The difference of their ages is the same every year, so Nathan is $23 - 17 = 6$ years older than Matt. This year their ages look like

Nathan	6
Matt	

$= 50$ years

If Matt was 6 years older, they would have the same age and the sum of their ages would be 56 years

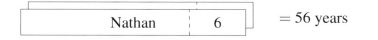

$= 56$ years

So, Nathan is $56 \div 2 = 28$ years old, and Matt is $28 - 6 = 22$ years old.

Example 4.4

David and Ed went to pick cherries. Together they picked 82 cherries. If David gave 4 cherries to Ed, they would have the same number of cherries. How many cherries did each of them pick?

Solution

We know they have 82 cherries together and they would have the same amount of cherries if David gave 4 cherries to Ed, so David has 8 more cherries than Ed:

We can see that if David had 8 less cherries, they would have $82 - 8 = 74$ cherries in total and then they would both have the same amount of cherries.

So, Ed has $74 \div 2 = 37$ cherries and David has $37 + 8 = 45$ cherries.

Example 4.5

Papa Bear and Baby Bear are peeling potatoes. Together they can peel 110 potatoes in 2 hours. If working separately for 5 hours, Papa Bear peels 25 more potatoes than Baby Bear. How many potatoes does each of them peel per hour?

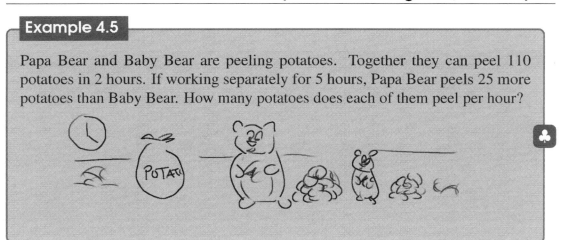

Solution

Since they can peel 110 potatoes working together in 2 hours, that means they can peel $110 \div 2 = 55$ potatoes working together in one hour. Let's look at the number of potatoes they can peel when they work separately in 5 hours:

				Papa Bear						

			Baby Bear				25		

So, in one hour Papa Bear peels $25 \div 5 = 5$ more potatoes than Baby Bear

PB	
BB	5

$= 55$ potatoes

If Baby Bear peeled 5 more potatoes each hour, they would peel together $55 + 5 = 60$ potatoes per hour.

PB

$= 60$ potatoes

So, Papa Bear peels $60 \div 2 = 30$ potatoes per hour and Baby Bear peels $30 - 5 = 25$ potatoes per hour.

Remark

Have you noticed a pattern already?

One thing we can see from this first few problems is that, when we are looking for *two numbers* and we know their sum <u>and</u> their difference, every time we have that

$$\text{Big Number} = (\text{Sum} + \text{Difference}) \div 2,$$

and that

$$\text{Small number} = (\text{Sum} - \text{Difference}) \div 2.$$

Try it out on the previous questions!

Example 4.6

In the orchard there are 144 trees, which are either apple trees or peach trees. If there were 12 fewer apple trees and 20 more peach trees, then the numbers of the two kinds of trees would have been the same. How many trees of each kind are there?

Solution

Let's take a look at how many trees of each kind we have:

If we indeed had 12 fewer apple trees and 20 more peach trees, we would have instead $144 - 12 + 20 = 152$ trees in total

 $= 152$ trees

and so we would have $152 \div 2 = 76$ of each. This means that we actually have $76 + 12 = 88$ apple trees and $76 - 20 = 56$ peach trees.

> ## Example 4.7
>
> Mary, Rose and Catherine are working on a school project and have some big pieces of ribbon of different sizes. If they glue together their pieces of ribbon they get a ribbon that is 840 centimeters long. Their teacher measured each of their ribbons and told them that Rose's ribbon is 130 centimeters longer than Mary's, and Catherine's is 220 centimeters longer than Rose's. Can you figure out how long is each ribbon?

Solution

Look at the pieces of Ribbon of each of the girls. Catherine has the longest piece of ribbon!

Let's cut the ribbons in smaller pieces. Take Catherine's ribbon and cut out of it 220 centimeters. So Catherine's ribbon is made of a piece as long as Rose's and one more that is 220 centimeters long.

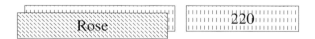

Now take those Rose pieces and cut 130 centimeters out of each. From each you get a piece that is as long as Mary's and an extra piece that is 130 long.

We know that all the pieces together would make up 840 centimeters, so those Mary pieces together will be

$$840 - 130 - 130 - 220 = 360$$

centimeters long, which means each Mary piece must be $360 \div 3 = 120$ centimeters long! And we also know that Rose's ribbon is $120 + 130 = 250$ centimeters long and Catherine's is $250 + 220 = 470$ centimeters long.

Example 4.8

The average price of a basketball, a soccer ball, and a volleyball is $36. The basketball is $10 more expensive than the volleyball, the soccer ball is $8 more expensive than the volleyball. How much is the soccer ball?

Solution

Since the average price of the balls is 36 dollars, the *sum* of their prices is $36 \times 3 = 108$ dollars.

Volleyball		
Basketball	10	= $108
Soccer ball	8	

If the Basketball was $10 cheaper and the Soccer ball was $8 cheaper, they would all cost the same as the Volleyball:

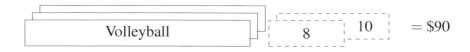

So, the price of *one* volleyball is $90 \div 3 = 30$ dollars. That means that the price of a Soccer ball is $30 + 8 = 38$ dollars.

Example 4.9

Captain Hook, Popeye, and Sinbad went out to the sea to hunt for treasure. They all found diamonds. The total number of diamonds found by Captain Hook and Popeye is 80. The total number of diamonds found by Popeye and Sinbad is 70. The total number of diamonds found by Captain Hook and Sinbad is 50. How many diamonds did each of them find?

Solution

The problem is telling us how many diamonds they got by pairs. Let's use bars to represent the number of diamonds each of them found.

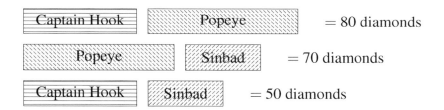

If we group all of these bars together, we have 2 bars of each and a total of 200 diamonds altogether.

So, if we had only one of each bar, we would have $200 \div 2 = 100$ diamonds. Note that if we get rid of two of the bars, we would know how many diamonds the third person found:

So, Sinbad found $100 - 80 = 20$ diamonds, Captain Hook found $100 - 70 = 30$ diamonds and Popeye found $100 - 50 = 100$ diamonds.

Example 4.10

John, Jack, and Jill have 159 marbles altogether. John has 2 more marbles than Jack, and if Jill gave 5 marbles to Jack, Jack would have the same number of marbles as Jill. How many marbles does each of them have? ♣

Solution

Let's take a look at the number of marbles each of them has:

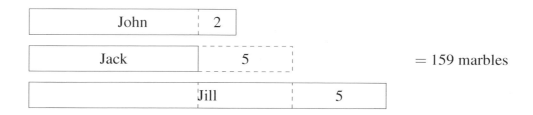

If Jill had 10 less marbles and John had 2 less marbles, they would all have the same number of marbles and they would have $159 - 2 - 10 = 147$ marbles in total

This means Jack has $147 \div 3 = 49$ marbles, John has $49 + 2 = 51$ marbles and Jill has $49 + 10 = 59$ marbles.

4.2 Quick Response Questions

Problem 4.1 There are 65 oranges in a grocery store arranged in two piles. If the first pile has 5 more oranges than the second pile, how many oranges does each pile have?

Problem 4.2 When Ian walks on stilts he is 17 feet tall. If he stands on the ground next to his stilts, the distance from his head to the top of the stilts is 5 feet. How tall is Ian?

Problem 4.3 Alek and Abby are collecting toy cars. Together they have 35 toy cars and Abby has 7 less than Alek. How many toy cars does Alek have?

Problem 4.4 Patrick and Joe ate all the cookies that their mom had just made. If her mom made 22 cookies and Joe ate 10 more cookies than Patrick. How many cookies did each of them eat?

Problem 4.5 Tim is 5 years old, and Allison is 3 years old. Their mom says she will take them to China when their combined age is 20. How old are they when they go to China?

Problem 4.6 Josh had 10 chocolate galleons that he shared with his friend Mike. Mike says that Josh ate 3 more galleons than him. Is that possible? How many did each of them eat?

Problem 4.7 One pig and one duck weigh 98 pounds. One pig and one turkey weigh 123 pounds. One duck and one turkey weigh 37 pounds. What is the weight of the turkey?

Problem 4.8 David likes collecting his bus tickets. Last week he took the bus 6 times less than this week and he gathered 34 bus tickets altogether. How many times did he ride the bus this week?

Problem 4.9 Quincey bought two boxes of pencils. One of the boxes was missing 4 pencils. If she has 62 pencils in total, how many pencils should each box have had?

Problem 4.10 There were two big stacks of boxes in a storage unit your dad works in. When your dad puts 3 boxes of the big stack into the small stack, they have the same height. If there are 16 boxes in total, how many boxes are there in each stack originally?

4.3 **Practice**

Problem 4.11 Four years ago, Curtis was 36 years older than Cassie. If this year their ages add up to 100 years, how old is Curtis?

Problem 4.12 Olive and Bluto collect stamps. They have 28 stamps in total. If Olive gives 6 stamps to Bluto, she still has 2 more stamps than Bluto. How many stamps do each of them have originally?

Problem 4.13 Again, the grocery store has two piles of oranges. There are 100 oranges in total. After 20 oranges are sold from the first pile, the two piles have the same number of oranges. How many oranges did each pile originally have?

Problem 4.14 Kyle and Lisa each bought some candies. If Kyle gave 4 candies to Lisa, they would both have the same number of candies. They pooled their candies together and counted, and there were 24 in total. How many candies did each of them originally have?

Problem 4.15 Patrick had three exams: history, math, and literature, and got a total score of 250 points. The score he got in the literature exam was 20 points less than the score he got in math, and was 10 points more than the score he got in the history exam. How many points did he receive in each exam?

Problem 4.16 Minsung and Patrick were playing Monopoly with some other friends. The number of properties that Patrick and Minsung have altogether is 24. If Patrick has 6 more properties than Minsung, how many properties does Minsung have?

Problem 4.17 Terrence has a small workshop where he builds chairs on his spare time. His son sometimes comes to help. When they work together they build 32 chairs in one weekend (2 days). If Terrence's son can build 4 less chairs than him in one day, how many chairs did each of them build over one weekend?

Problem 4.18 Adam, Bob and Chris have $875 altogether. Adam has $250 more than Bob, and Bob has $125 more than Chris. How much money do each of them have?

Problem 4.19 We planted 121 different plants in the school's garden. Unfortunately some of the plants got eaten by some squirrels. The number of plants that survived is 61 more than the number of plants eaten by the squirrels. How many of the plants did the squirrels eat?

Problem 4.20 In a school there are totally 108 students in the third, fourth, and fifth grades. The third grade has 11 fewer students than the fourth grade, and the fourth grade has 16 more students than the fifth grade. How many students are there in each grade?

Problem 4.21 Hercules bought 5 swords and 2 spears for 304 drachmas (drachma is the currency of Greece). A spear cost 9 drachmas less than a sword. How much does a sword cost?

Problem 4.22 Frank took three quizzes, with a total score of 148 points. His quiz A score is 21 points higher than his quiz B score, and his quiz C score is 32 points lower than his quiz A score. What is his score for each of the quizzes?

Problem 4.23 Old Sealegs Austin, Landlubber Matthew, and Cap'n Poopdeck went to search for treasure. They found 130 gold coins in total. Old Sealegs Austin found 7 more gold coins than Landlubber Matthew, and if Old Sealegs Austin and Landlubber Matthew put their gold coins together they would have 24 more than Cap'n Poopdeck's. How many gold coins did each of them find?

Problem 4.24 Of the 80 levels I completed on my video game, it took two tries to complete some levels and only one try to complete other levels. The number of levels which only took one try is 20 more than the number of levels which took two tries. How many levels only took one try?

Problem 4.25 I tried to squirt Dad with a hose 101 times last summer. I missed 31 more times than the number of times I got him wet. How many times did I miss trying to squirt Dad?

Problem 4.26 Myles, Mya and Myron entered a team competition where they each had to run 400 meters and add up their times to get their team score. Myles was 10 seconds faster than Myron, and Mya was 4 seconds faster than Myron. If their total time was 181 seconds, how long did it take Myron to complete the course?

Problem 4.27 The twins Walker and Cooper are coming to a Halloween party dressed up as a gigantic ghost (so, Walker will be standing on Cooper's shoulders and then they will cover themselves with a huge blanket). They were so tall that Walker hit his head with a lamp that was 11 feet high. If their heads are 1 foot long, how tall are the twins?

Problem 4.28 Cameron, Mandy and Taylor decided to put all their savings together to buy the new video game console that all the kids were playing. They had enough money to buy the video game console that cost $300 and an extra controller that cost $20. If Taylor had $20 less than Mandy, and Mandy had $50 more than Cameron, how much money did each of them have?

Problem 4.29 Wilson watched the movie Cast Away and got obsessed with Wilson, the volleyball, because they have the same name. He visited some sports stores and every time he bought more volleyballs than in the previous store. He bought in total 19 volleyballs in three different sports stores. In the second store he bought 2 more balls than in the first store and in the third store he bought 3 more than in the second store. How many balls did he buy in the first store?

Problem 4.30 Cody likes separating his M&M's by color before eating them. He has now only blue and green M&M's left. Before he ate 3 blue M&M's and 7 green M&M's, he had 50 M&M's. Now he has 2 more green M&M's than blue M&M's. How many green M&M's does he have left?

Problem 4.31 Rosemary loves cooking with fresh herbs so she has some pots with thyme and rosemary, which is her favorite, in her backyard. If she had 2 more thyme pots and 4 less rosemary pots, she would still have 10 more rosemary pots than thyme pots. If she has 24 pots in total, how many pots of each plant does she have?

Problem 4.32 Claude needed to weigh his dog Ty, but his other two dogs, Jy and Xy, were giving him problems. Every time he placed Ty on the scale at least one of the other dogs stepped on it too. If Ty and Jy weighed 103 pounds, Ty and Xy weighed 112 pounds, and the three of them together weighed 175 pounds, how heavy is Ty?

Problem 4.33 Two hungry ant-eaters were feeding from a small anthill. If one of the ant-eaters eats 150 more ants than the other and they ate a total of 2300 ants, how many did each of them eat?

Problem 4.34 Mom and I went to the store to buy some utensils for a big dinner party we will be having over the weekend. We already had some so we didn't need to buy whole sets. We bought spoons, forks and knives, 70 in total. If we bought 10 more spoons than forks and 5 less spoons than knives, how many of each utensil did we buy?

Problem 4.35 Rita needed to buy some cat food. Her cat, Melrose, is picky when she eats so she needs to buy different kinds of food. In total Rita bought 38 cans of food. She bought 4 more cans of "Savory Shreds" than "Classic Paté", and 10 more cans of "Prime Filets" than "Classic Paté". How many cans of each kind did she buy?

5. Counting Without Fingers

It's the big day for your favorite cousin. She is getting married! And, as you are her favorite cousin as well, she has come up with an important task for you: You have to make sure to place the numbers on the tables for the reception dinner so that all the guests know where they should sit. You know there will be 150 tables at the reception dinner and you will need to use a separate sticker for each of the digits that you will need. This seems like an easy task but, before you start, how many stickers are you going to need?

We all know how to count: "one, two, three, four, ...", but counting too high can be boring. Its much more fun to count: "one, two, skip a few, ninety-nine, a hundred". But how do we know how many numbers to skip when we "skip a few"? In this chapter we'll learn ways to organize our counting so we can count things quickly and without mistakes!

The concepts introduced in this chapter directly correspond to Common Core Math Standards as shown in the following table.

3rd Grade	3.OA.1, 3.OA.8
4th Grade	4.OA.2, 4.OA.3

In addition to the standards above, problems and concepts in this section will help strengthen understanding of the following domains.

3rd Grade	3.OA, 3.NBT, 3.MD
4th Grade	4.OA, 4.NBT, 4.MD
5th Grade	5.OA, 5.NBT, 5.MD

5.1 Example Questions

Example 5.1

Your cousin asked you for help setting up the numbers for the tables at her wedding. She gave you lots of stickers with all the available digits. If you want to label 150 ♣ tables, how many stickers are you going to use?

Solution

For the first 9 tables you will only use one sticker per table since those are 1-digit numbers. The 2-digit numbers go from 10 to 99 and there are exactly 90 of them. Since 150 is bigger than 99 you will need all of those, and so you will use $90 \times 2 = 180$ stickers for the 2-digit numbered tables. We will not need all the 3-digit numbers, since they go from 100 to 999 and we only need up to 150. So we just need to count how many numbers are there between 100 and 150, including the 100 and 150. From 101 to 150 there are 50 numbers, since we also need 100, that means we have 51 3-digit numbers. So we will need $51 \times 3 = 153$ stickers for the 3 digit numbered tables. In total we will need

$$9 + 180 + 153 = 342$$

stickers for all the tables.

Example 5.2

It seems that your cousin really likes the number 2, so now she's asking you to replace all the 2s in the labels with a special number 2 sticker. How many stickers will you have to replace in total?

Solution

This time, we'll consider the numbers place-by-place. The digit 2 will appear in the ones place one time in every group of 10 consecutive numbers. For the labels of tables 1 through 150 you need exactly $150 \div 10 = 15$ such groups, so 2 appears in the tens place 15 times: 2, 12, 22, 32, ..., 122, 132, 142.

The digit 2 will appear in the tens place ten times in every group of 100 consecutive numbers. In the numbers between 1 and 100 this will only happen with the numbers that start with 2: 20, 21, 22, ..., 28 and 29, so we have 10 so far. From 101 to 150 this happens again 10 times since we are using all the numbers from 120 to 129.

The digit 2 will not be used in the hundreds place since the biggest number you need is 150 and the smallest number with a 2 in the hundreds place is 200.

So, in total you have to replace $15 + 20 = 35$ stickers that have the digit 2.

Remark

Counting this way we did not have to worry about numbers using the digit 2 more than once, like 22 or 122, since we had them in both lists: the numbers that use 2 in the ones place and the numbers that use 2 in the tens place.

Example 5.3

Timmy is getting ready to go to a party and is having trouble deciding what to wear. He has 3 different pairs of jeans, 8 different shirts and 2 different pairs of shoes. In how many different ways can he get dressed?

Solution

Note that for each choice of jeans he can choose whatever shirt he wants, and for each choice of jeans and shirt he can choose any pair of shoes. Since he has 3 pairs of jeans, 8 shirts and 2 pairs of shoes, he can get dressed in $3 \times 8 \times 2 = 48$ different ways.

Example 5.4

How many whole numbers between 99 and 999 have exactly one 0?

Solution

Let's take a look at the numbers in the list from 100 to 199. They all have a 1 as the first digit. There are only 10 numbers in there that have 0 in the tens digit: 100, 101, 102, ..., 108 and 109, but 100 has two 0s, so we have 9 numbers with exactly one 0 so far. The rest of the numbers that will have 0 will have it in the tens digit: 110, 120, 130, ..., 190. So, from 100 to 199 we have $9 + 9 = 18$ numbers that have exactly one 0. Note that we can count in the same way the numbers that have something other than 1 in the hundreds digit. Since we have 9 possible digits we can choose for the hundreds digit, we will have in total $18 \times 9 = 162$ numbers with exactly one 0.

Example 5.5

You are using a vending machine that only accepts bills and does not give change back. You need to pay $17 and you have only $5 bills and $2 bills. In how many ways can you choose how to pay?

Solution

Observe first that 17 is an odd number. That means you will need to have an odd number of $5 bills, otherwise you would end up with an even amount of money. If you use one $5 bill, you will need to use six $2 bills to make up for the remaining $12. If you use three $5 bills, you will need only one more $2 bill to have $17 total. Thus, there are only 2 different ways that you can choose what bills to use to pay the $17.

Example 5.6

Ali, Bonnie, Carlo and Dianna are going to drive together to a nearby theme park. The car they are using has four seats: one driver's seat, one front passenger seat and two back seats. Bonnie and Carlo are the only two who can drive the car. How many possible seating arrangements are there?

Solution

Let's pick the people that will seat in each of the seats of the car in order. We will start with the driver's seat. Since only Bonnie and Carlo can drive, we have 2 choices for the driver's seat. For the front passenger seat we have 3 choices, since we already chose one person to drive. For the left back seat we have 2 choices left. For the last seat we only have 1 choice since all other seats have already been selected. In total we have

$$2 \times 3 \times 2 \times 1 = 12$$

ways of arranging the passengers in the car.

Example 5.7

Ms. Hamilton's 8th grade class wants to participate in the annual three-person-team basketball tournament. Lance, Sally, Joy and Fred are chosen for the team. In how many ways can the three starters be chosen?

Solution

Note that when you choose 3 students to play on the tournament from a group of 4, you are choosing at the same time 1 student that will not be a starter player. There are clearly 4 ways to choose who is not starting the game, so there are 4 ways to choose the first three starters for the basketball tournament.

Example 5.8

If 50 people are in a room and everyone shakes everyone else's hand once, how many handshakes were exchanged in the room?

Solution 1

Let's pretend that the room is empty and that when someone new enters the room, that person shakes hands with everyone already in the room. The first person to enter the room will shake hands with no one because the room was empty. The second person will shake hands only with 1 person. The third person will shake hands with 2 people... Notice that every time a new person arrives into the room, the next person will have one more hand to shake. The last person to arrive will have to shake everyone else's hands, that is, will shake 49 hands. So, we will have in total $1 + 2 + 3 + 4 + \cdots + 49 = 1225$ handshakes.

Remark

Do not worry right now how we were able to add up all those numbers so quick, we will learn how to do that in one of the following chapters of the book!

Solution 2

We can also count everything in one single step. Notice that after all handshakes were exchanged, each person will have shaken the hands of other 49 people. However, we want to be careful: when counting like this we are counting every handshake twice. So, in total we have $50 \times 49 \div 2 = 1225$ handshakes.

Example 5.9

How many 2-digit counting numbers have *at least* one 7 as a digit? ♣

Solution

First of all, note that the question is using the words *at least*. This means we want the numbers that have exactly one 7 and also the numbers that have exactly two 7s. We have 9 2-digit numbers that end in 7: 17, 27, ..., 97; and we have 10 2-digit numbers that start with 7: 70, 71, 72, ..., 78, 79. We want to be careful though, because the number 77 starts with 7 and also ends in 7 which means we have it in both of our lists already. So, in total we have $9 + 10 - 1 = 18$ numbers that have at least one 7.

Example 5.10

Tyler entered a buffet line in which he has to choose one kind of meat, two different vegetables and one dessert. Available meats are beef, chicken and pork; available vegetables are baked beans, corn, potatoes and tomatoes; and for dessert he can choose from brownies, chocolate cake, chocolate pudding and ice cream. If the order of the food items is not important, how many different meals might he choose? ♣

Solution

There are 3 kinds of meat available, 4 kinds of vegetables and 4 kinds of dessert to choose from. Let's deal first with choosing the vegetables: If Tyler had to choose them

---

in a particular order, he would have $4 \times 3 = 12$ different possible ways of choosing them; since he doesn't care which one he chooses first, we need to divide 12 by 2 (there are 2 ways of swapping the order of the vegetables once they are chosen), so he has 6 ways of choosing the vegetables. For the meat and dessert, we don't need to worry as much because he is just picking one of each. So, there are 3 ways of choosing the meat, 6 ways of choosing the vegetables and 4 ways of choosing the dessert. This means that in total Tyler has $3 \times 6 \times 4 = 72$ ways of choosing his meal at the buffet table.

5.2 Quick Response Questions

Problem 5.1 Find the number of 2-digit positive integers whose digits sum a total of 7.

Problem 5.2 If 5 people are in a room and everyone shakes everyone else's hand once, how many handshakes were exchanged in the room?

Problem 5.3 How many 4-digit numbers can you form with the digits 3, 7, 8 and 9 if you can use each digit only once?

Problem 5.4 How many 3-digit numbers can you form with the digits 1, 5 and 7? You can repeat digits.

Problem 5.5 In how many ways can you choose what jacket and hat to wear if you have 4 jackets, 7 baseball caps, and 2 beanie hats? (You will only wear one hat at a time!)

Problem 5.6 When writing the integers 120 through 180, how many times is the digit 3 written?

Problem 5.7 Jamie flips a coin 3 times. How many possible ways can Jamie get exactly 2 heads?

Problem 5.8 How many different sums can you get adding two or more of the numbers 1, 3, 4 and 7?

Problem 5.9 How many different 4-digit numbers can be formed by rearranging the four digits in 2004?

Problem 5.10 Of the 30 letters shown below, how many of them are *O*s?

IOOOOOOOOOOOOOOOOOOOOOOOOOOOOI

5.3 Practice

Problem 5.11 There are 3 different kinds of chili, 2 kinds of tomatoes, 3 kinds of onions and 2 kinds of avocado at the grocery store. You want to make some guacamole choosing one of each kind of these ingredients. In how many different ways can you choose the ingredients?

Problem 5.12 Thomas has a pen that he can only use for writing numbers and will run out of ink right after he writes the digit 7 a total of 10 times. He wants to write the counting numbers starting from 11. What is the last number Thomas will be able to write?

Problem 5.13 Thomas bought a new and improved pen that allows him to write up to 20 times the digit 7. This time he starts writing the counting numbers from 100. What is the last number he will be able to write?

Problem 5.14 How many 4-digit even numbers can you form with without repeating digits and without using the digit 0?

Problem 5.15 Ricky has to choose 8 of his 10 favorite toys to bring on a long road trip. In how many ways can he choose what toys to bring?

Problem 5.16 You are helping your Mom to make a fruit arrangement. How many different arrangements can you make if you must choose one from 3 kinds of pears, one from 6 kinds of berries, one from 4 kinds of apples, and one from 7 kinds of nuts?

Problem 5.17 Three friends have a total of 6 identical pencils, and each one has at least one pencil. In how many ways can this happen?

Problem 5.18 The Little Twelve Basketball Conference has two divisions, with six teams in each division. Each team plays each of the other teams in their own division twice and every team in the other division once. How many conference games are scheduled?

Problem 5.19 Henry's Hamburger Heaven offers its hamburgers with the following condiments: tomato, lettuce, pickles, and cheese. A customer can choose one, or two meat patties, and any collection of condiments. How many different hamburgers can be ordered?

Problem 5.20 If you have 3 nickels, 5 pennies and 1 dime. How many distinct ways can you give someone 20¢?

Problem 5.21 How many 3-digit numbers can be formed using 3, 5, and 0? Digits may not be repeated.

Problem 5.22 How many 4-digit numbers can be formed using the digits 0, 1, 2, 3, 4 and 5? Digits may not be repeated.

Problem 5.23 Four friends, Ivory, Katie, Nick and Peter, go to the mysterious pyramid to find treasures. They tried the secret code *Open Sesame* and the gate did not open, but instructions were displayed: "The secret code has to be formed using only and all of your first name initials one time". They tried a lot of different codes and they got the door open using the last code they could try, how many different codes did they check?

Problem 5.24 Ursula is making up 5-letter words with some scrabble tiles she found under her sofa. She found the letters *J, A, C, K* and *Y*. How many different "words" can she make if she wants the third letter not to be *Y*?

Problem 5.25 You are setting up the passcode for your new phone. You have to choose 1 letter followed by 3 digits. How many codes can you make if you only want to use odd digits and you do not want to use vowels?

Problem 5.26 At the DMV they have a new rule about license plates. Before they allowed only one letter followed by 5 digits, now they allow 2 letters followed by 4 digits. How many more license plates can they have with this new rule?

Problem 5.27 A grocer stacks oranges in a pyramid-like stack whose rectangular base is 5 oranges by 8 oranges. Each orange above the first level rests in a pocket formed by four oranges in the level below. The stack is completed by a single row of oranges. How many oranges are there in the stack?

Problem 5.28 There are seven students and seven chairs lined up in a room. In how many ways can the students be seated in the room?

Problem 5.29 In how many ways can the students from the previous problem be seated if we arrange the chairs in a circle and the oldest person has to be seated in the chair that is facing the door?

Problem 5.30 When your family of 8 people gets together for Christmas, you all sit together around a rectangular table that has 8 chairs. Your grandfather always sits at the head of the table and your grandma sits to his side. In how many ways can you all choose to sit at the table?

Problem 5.31 John is playing Monopoly with his friends. If he has four $M\$3$ bills and three $M\$5$ bills, how many different amounts of money could he pay to someone by using one or more bills?

Problem 5.32 Today's Chef's Specials were served on 23 plates of different colors: 8 red plates, 6 green plates, 5 white plates, and 4 blue plates. How many of today's Chef's Specials were not served on blue plates?

Problem 5.33 Pat will select three cookies from a tray containing only chocolate chip, oatmeal, and peanut butter cookies. There are at least three of each of these three kinds of cookies on the tray. How many different assortments of three cookies can be selected?

Problem 5.34 Your mom wants you to go and buy some lottery tickets for her. She told you to buy "all the tickets that end in 6 and start with 894 with no repeated digits". If each lottery ticket has a 7-digit number and they never use the digit 1, how many tickets can you buy?

Problem 5.35 Daniel is about to buy a new car. He already knows the model of the car he wants to buy but he can also choose the color of the exterior and the color of the seats, as well as extra features. There are 7 exterior colors available, 2 colors for the seats, and he can choose to add GPS navigation system, satellite radio and heated seats. In how many ways can Daniel configure the car he wants to buy? (Note that he may add more than one extra feature or may not want to add any extra features at all).

6. Figurate Numbers

Lawrence came across some billiard balls in his dad's game room at home. He has no idea how to play billiards, since he is just a kid, but still he has seen that whenever they are about to start a game they set up a triangle with the balls on the table.

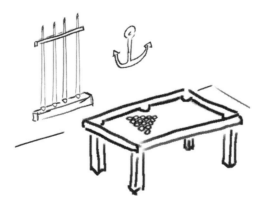

He then decided to play with the balls by making some different triangles, but he is having trouble figuring out how many balls he will need for a big triangle that has 10 balls per side instead of the usual 5. How many balls would he need?

Everywhere we look we see patterns all around us. Being able to recognize patterns and knowing common types of patterns is helpful not only in math but in other subjects as well. Patterns can involve numbers, shapes, or even combine the two! You even get to practice your drawing skills while learning patterns!

The concepts introduced in this chapter directly correspond to Common Core Math Standards as shown in the following table.

3rd Grade	3.OA.3, 3.OA.9
4th Grade	4.OA.2, 4.OA.3, 4.OA.5
5th Grade	5.OA.3

In addition to the standards above, problems and concepts in this section will help strengthen understanding of the following domains.

3rd Grade	3.OA, 3.NBT, 3.G
4th Grade	4.OA, 4.NBT
5th Grade	5.OA, 5.NBT

6.1   Example Questions

Example 6.1

Let's help Lawrence pick the number of balls he needs. How many billiard balls would he need to build a triangle with 10 balls on each side?

Solution

Let's start by making smaller triangles. Perhaps we can figure out a way to find out the total number of balls without having to draw all the balls. Triangles with 1, 2, 3 and 4, balls per side would look like

Actually, in problems like this it is always helpful to try to spot a pattern in smaller versions of the problem. When the information is presented in an organized way it is a lot easier to do so. Look at the table below:

# of balls per side	1	2	3	4	5	$\cdots$
# of new balls added	1	2	3	4	5	$\cdots$
Total # of balls	1	3	6	10	15	$\cdots$

Note that each time we add as many balls as the number of balls that we will have on the side of the triangle. This means that a triangle with 10 balls per side could be constructed by starting with a 1-ball "triangle", then adding 2 balls, then 3 more, and so on, for a total of

$$1 + 2 + 3 + 4 + 5 + 6 + 7 + 8 + 9 + 10 = 55$$

balls.

Remark

The total number of balls we need to build these triangles are called *triangular numbers*: 1, 3, 6, 10, 15, 21, 28, 36, ...

Example 6.2

Find the 12th term of the sequence 2, 4, 7, 11, 16, ... ♣

Solution

Note that in this sequence of numbers we always have 1 more than the triangular numbers:

$$1 + 1, 3 + 1, 6 + 1, 10 + 1, 15 + 1, \ldots$$

So, the 12th number that will appear on this list is going to be 1 more than the 12th triangular number. In a previous problem we found up to the 10th triangular number, 55. The next two are $55 + 11 = 66$ and $66 + 12 = 78$, so the number we are looking for is $78 + 1 = 79$

Example 6.3

Dary, the ant, came across a stack of oranges arranged like a triangle with 5 oranges per side and wanted to get to the very top. Dary wants to get there as soon as possible, so she will only move up.

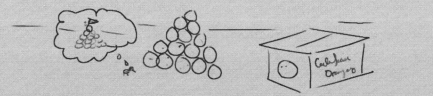

(a) If Dary starts climbing from the first orange in the bottom layer, in how many different ways can she get to the top?

(b) What if she starts from the third orange in the bottom layer?

Solution to Part (a)

The first orange in the bottom layer has only one orange directly above it, so Dary has no other choice when she goes to the next layer than to just go up to that orange.

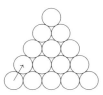

The same thing will happen each time she goes up one layer since every time there is just one orange resting on top of the orange she is standing in.

So Dary has only one possible way of climbing to the top of the pyramid.

Solution to Part (b)

This time Dary has more choices when she moves up since she could start by going up to the left or to the right. These are some paths she could follow:

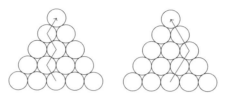

To figure out the answer to the problem it might be easier to start from the top and go down. Let's pretend Dary starts at the top and goes down to each of the oranges in the middle layers, each time writing on each orange the number of ways she could get there. She has only 1 way to get to each of the oranges in layer 2 since she can only come down from the orange on the top.

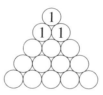

When she comes to the third layer, there is only one path she can follow to the first and third oranges, but there are two paths she can follow to go to the orange in the middle.

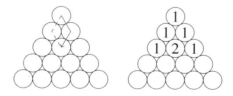

As she goes to the fourth layer she has way more options. Look at the second orange in the layer. She can come down from the orange that has a 1, or from the orange that has a 2. This means that in total, she can come down to the second orange in the fourth layer in 3 different ways. Same thing goes for the third orange.

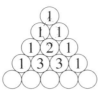

And to the bottom layer, she can follow the same pattern. Notice that this way of looking at the paths, we just need to look at the two oranges on top of the orange we are looking at and add the numbers they have on them:

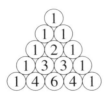

So, there are 6 possible ways that she can go from the orange in the middle of the bottom layer to the very top of the pyramid.

Remark

The triangle we ended up setting up to help Dary figure out the number of paths is called *Pascal's triangle*. It can have as many layers as we want. To set if up we will start with a 1 on top, and also 1s at the beginning and end of each layer. All the numbers in between the 1s in each layer will be obtained by adding up the two numbers on top of it.

In many areas of mathematics it is useful to set it up to figure out answers to some problems in a quick way. Also, there are some beautiful patterns that come with Pascal's triangle. You will be able to see some of them in a couple of the practice problems ahead.

Example 6.4

If today were Friday, what day of the week would it be 75 days from now?

Solution

The days of the week always come in the same order, and they repeat over and over again every 7 days. This is an example of a pattern we can identify easily. We will not need to list all 75 days after today if we take advantage of this. Since the days of the week repeat every 7 days, we can say that it will be Friday again 7 days from now, 14 days from now, 21 days from now, and so on. If we follow this pattern we can see that it will also be Friday 70 days from now, 75 days from now will be the same day of the week as 5 days after Friday, so it will be Wednesday.

Remark

When you come across repeating patterns you can think as if you were stuck in a loop, where you start counting again whenever the pattern starts repeating again. So you could think of

Fri, Sat, Sun, Mon, Tue, Wed, Thu, Fri, Sat, ...

the same way as

$$1, 2, 3, 4, 5, 6, 7, 1, 2, \ldots$$

Example 6.5

The word *MATH* is repeatedly written producing the pattern

$$MATHMATHMATHM\ldots$$

If the pattern is continued, what letter will occur in the 2017th position?

Solution

The word *MATH* has four letters, so the letters in the pattern with repeat over and over again every four letters. To see how many times we can fit the whole word *MATH* writing no more than 2017 letters, we can divide 2017 by 4. When we divide $2017 \div 4$ we get a quotient of 504 and a remainder of 1. That means that if we were to write all the letters in the sequence up to the 2017th letter, we would write the word *MATH* 504 times followed by *one* letter *M*. Therefore, the answer is *M*.

Example 6.6

What is the units digit of 3^{1002}?

Solution

Remember: 3^{1002} means that we will multiply 3 by itself 1002 times

$$\underbrace{3 \times 3 \times 3 \times \cdots \times 3}_{1002 \text{ times}}$$

Since we only want to know the last digit of the number, we might not need to find the actual number, which is quite big.

Let's take a look at the first few powers of 3:

power	number	last digit
3^1	3	3
3^2	9	9
3^3	27	7
3^4	81	1
3^5	243	3
3^6	729	9
3^7	2187	7
3^8	6561	1

Have you noticed a pattern already? If we look at the last digit column in the table, we can see that they form a pattern that repeats every *four* numbers. So we will find the answer to the problem if we respond this question: *What is the 1002nd number in the repeating sequence 3, 9, 7, 1, 3, 9, 7, 1, ...?* Note that to answer this question we would need to do the same as we did in the previous problem with the repeating *MATHMATH...* pattern. Since we get a remainder of 2 when we divide 1002 by 4, we want the second number in the repeating list: 9

Example 6.7

Debra is counting beans out loud to make groups of five beans for an art project. Her little brother was paying attention to what she was saying: *"one, two, three, four, one, one, two, three, four, two, one, two, three, ..."*. If Debra had 64 beans, how many times did her little brother hear her say the word *three*?

Solution

Let's figure out first what is the pattern Debra is using to count. If we look closely at the first numbers she counts out loud, she never reaches the number five, but instead says out loud how many groups of 5 beans she has counted so far:

				# of groups
				↓
one,	two,	three,	four,	one,
one,	two,	three,	four,	two,
one,	two,	three,	...	

The first four words in every group of 5 are always the same and the only one that changes is the fifth one. So this is almost a pattern that repeats itself every five terms. Since she has 64 beans, she will be able to count $60 \div 5 = 12$ groups of 5 and will have 4 left over. The last words she would say are

...,four, twelve, one, two, three, four.

So her brother will hear the word *three* in each of the 12 complete groups of 5, one time when she counts 3 groups, and one time within the last 4 beans she counts. That is, $12 + 1 + 1 = 14$ times in total.

Example 6.8

What is the value of

$$-3 + 6 - 9 + 12 - 15 + 18 - 21 + 24 - \cdots - 99 + 102 \quad ?$$

This is a long row of sums and differences that seems to follow a pattern. Note that all the numbers are multiples of 3 and the $-$ and $+$ signs alternate. Let's take a look at shorter versions of this problem to see if we can spot a pattern:

	Total
-3	-3
$-3+6$	3
$-3+6-9$	-6
$-3+6-9+12$	6
$-3+6-9+12-15$	-9
$-3+6-9+12-15+18$	9
$-3+6-9+12-15+18-21$	-12
$-3+6-9+12-15+18-21+24$	12

Did you notice that whenever the last number we have is positive the total is exactly half as much as that number? In the original problem we want to figure out, the last term is a $+102$ so the total must be $102 \div 2 = 51$.

Example 6.9

Kori is bored at her mom's office and started playing with some squared sticky notes she found on her desk. She made some figures in one of the walls that look somehow like this:

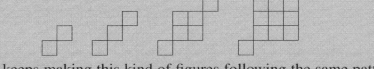

(a) If she keeps making this kind of figures following the same pattern, how many sticky notes will she need for the 6th figure?

(b) How many sticky notes would she need for the 51st figure?

Solution to Part (a)

We can see that she starts the first figure with just two sticky notes and then for for the next ones she is first making a square with the sticky notes and then adding two extra sticky notes in opposite corners of the square.

The next two figures she would make are

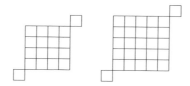

so we can see that the 6th figure would need 27 sticky notes.

Solution to Part (b)

If we want to draw the figure and count how many sticky notes it has it is going to take forever. Instead, let's use what we have found out so far from her pattern:

Figure	Size of big square	# of sticky notes
1	—	2
2	1×1	3
3	2×2	6
4	3×3	11
5	4×4	17
6	5×5	27

Note that the size of the side of the big square is one less than the figure number. So, the 51st picture will be a 50×50 square and two sticky notes on the corners, so Kori would need $50 \times 50 + 2 = 2502$ sticky notes for that figure.

Example 6.10

Find the missing numbers in the following pattern.

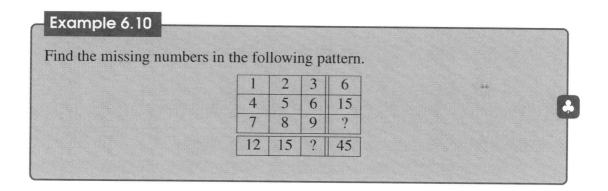

Look at the first two rows. Note that

$$1+2+3=6$$

and

$$4+5+6=15.$$

We can see the same thing happening in the first two columns:

$$1+4+7=12$$

and

$$2+5+8=15.$$

This points to the pattern being "the last row/column is the sum of the first three". Let's try adding the first three numbers in the third row

$$7+8+9=24$$

and the first three numbers in the third column

$$3+6+9=18.$$

Note that the pattern still works fine if we add up the first three numbers in the last row

$$12+15+18=45$$

and in the last column

$$6+15+24=45$$

so the missing numbers are indeed 18 and 24.

1	2	3	6
4	5	6	15
7	8	9	**24**
12	15	**18**	45

6.2 Quick Response Questions

Problem 6.1 Identify a pattern and figure out the next number in each sequence.

   (a) 1, 4, 9, 16, 25, ...

   (b) 1, 5, 14, 30, 55, ...

   (c) 2, 6, 12, 20, 30, ...

Problem 6.2 Consider the number sequence 6, 1, 3, 6, 1, 3, 6, ... What is the 20th number in the sequence?

Problem 6.3 What is the units digit of 2^{2017}?

Problem 6.4 What is the next figure in the sequence?

(a) $\triangle, \square, \triangle, \square, \ldots$

(b) $\star\triangleleft, \star\triangleright, \star\triangleleft, \ldots$

(c) $\square, \triangle\square\square, \triangle\triangle\square\square\square, \ldots$

Problem 6.5 I'm wearing a hat that I got 17 days before Tuesday. What day was that?

Problem 6.6 Identify a pattern in each sequence of triangles and draw the next triangle in the sequence.

(a)  $\blacktriangle, \blacktriangle, \blacktriangle, \triangle, \ldots$

(b) $\triangle$, $\blacktriangledown$, $\triangle$, $\blacktriangledown$, ...

Problem 6.7 Lindy is about to start an internship on September 1st. If the internship lasts for 20 months, what is the last month she will be working there?

Problem 6.8 I got my new bike six days after Tuesday. What day of the week did I get my bike?

Problem 6.9 Suppose yesterday was Monday. If we start from today and count the days, what day would it be on the 18th day?

Problem 6.10 The inspector came to Roy's restaurant on Monday. He said that he would be coming back 8 business days later. What day of the week is he coming back? (Business days are Monday through Friday).

6.3 Practice

Problem 6.11 What is the 10th number in the following sequences?

(a) $1, 2, 2, 3, 3, 3, 4, \ldots$

(b) $22, 33, 333, 44, 444, 4444, 55, \ldots$

(c) $0, 1, 10, 11, 100, 101, 110, 111, \ldots$

Problem 6.12 Answer the first few questions and try to spot a pattern to answer the last question. Note: three or more points are noncollinear if you cannot draw a *straight line* that goes through more than two of them at the same time.

(a) How many straight lines can be drawn between 2 points?

(b) How many lines can be drawn between 3 noncollinear points?

(c) How many lines can be drawn between 4 noncollinear points?

(d) How many lines can be drawn between 20 noncollinear points?

Problem 6.13 Clive is stacking boxes at work. He is super meticulous and tries to follow a pattern every time he adds more boxes to the existing stack of boxes he has. He started with no boxes. His stack of boxes looked like this the first four times he added boxes to it:

(a) What is the pattern that Clive is following?

(b) How many boxes will he have in his stack after the 6th time he adds boxes to it?

(c) How many boxes does he need to add to the 6th stack to get the 7th stack?

Problem 6.14 Look again at the previous problem and fill in the following table. Then answer the following questions.

Stack number	# of boxes added	total # of boxes
1	1	1
2	3	4
3	5	
4		
5		
6		
7		

(a) What pattern do you notice in the number of boxes that are being added each time the stack grows?

(b) What pattern do you notice in the total number of boxes in each stack?

(c) What is the sum of the first 100 odd counting numbers?

(d) Find the sum $11 + 13 + 15 + \cdots + 97 + 99$

Problem 6.15 What is the next figure in the sequence?

(a) ♡◇, ♡◇♡◇, ♡◇♡◇♡◇, ♡◇♡◇♡◇♡◇, . . .

(b) ⋆★⋆, ⋆★★⋆, ⋆★★★⋆, ⋆★★★★⋆, . . .

(c) ←, →, ⇇, ⇉, ←⇇, ⇉→, ⇇⇇, ⇉⇉, . . .

Problem 6.16 There is a number sequence

$$3, 9, 7, 4, 7, 2, 5, 5, 3, 9, 7, 4, 7, 2, 5, 5, 3, 9, 7, 4, \ldots$$

What is the 2017th term in the sequence?

Problem 6.17 Fill out the missing term in the figure sequence.

(a) ◁▷⋆∪,   ▷⋆∪◁,   _____,   ∪◁▷⋆,   ◁▷⋆∪,   ...

(b) ↑,   ↓↑↓,   _____,   ↓↑↓↑↓↑↓,   ↑↓↑↓↑↓↑↓↑,   ...

(c) ⋆⋆,   ⋆‖⋆,   ⋆|⋆⋆|⋆,   _____,   ⋆|⋆|⋆⋆|⋆|⋆,   ...

Problem 6.18 Kori was back at her mom's office. She had so much fun making patterns with sticky notes last time that she went on with it again. This time her first patterns looked like this:

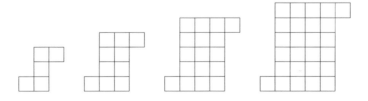

(a) How many sticky notes would she need for her 6th figure?

(b) Write a description that could be used to determine the shape and the total number of sticky notes in figure 6. Your description should be clear enough so that another person could read it and use it to think about another figure.

(c) How many sticky notes would she need for the 30th figure?

Problem 6.19 Find the missing numbers in the following pattern:

1	2	4	1
2	?	7	0
6	7	14	1
?	0	3	0

Problem 6.20 Find the value of $99 - 98 + 97 - 96 + \cdots + 3 - 2 + 1$

Problem 6.21 What is the next figure in the sequence?

(a) △□, △△□, △△△△□□, △△△△△△□□, △△△△△△△□□□□, ...

(b) ↓, ↑↑↓↑↑, ↓↓↓↑↑↓↑↑↓↓↓, ↑↑↑↑↓↓↓↑↑↓↑↑↓↓↓↑↑↑↑, ...

(c) ◁▷, ◁◁▷◁▷▷, ◁◁◁▷◁▷▷◁◁▷◁▷▷▷, ...

Problem 6.22 Chandler came up with a rule to write a sequence of numbers: if the last number he wrote is even, he divides it by 2 and that is the next number in the sequence; if the last number he wrote is odd, he adds 1 to it and then divides it by 2 to come up with the next number. He decided that the first number would be 250. What is the 15th number in his sequence of numbers?

Problem 6.23 Consider the following number sequence:

$$101, 102, 103, 100, 105, 98, 107, \ldots$$

(a) What is the 10th term in the sequence?

(b) What is the 15th term in the sequence?

(c) Try to find more than one way to describe the pattern the numbers are following.

Problem 6.24 Tom was bored while waiting for his mom to come to pick him up from school. While he waited he repeatedly wrote "TOMISAWESOMESAUCE" in a small piece of paper (leaving no spaces in between each time he wrote the phrase). He had just enough space to write 230 letters in the piece of paper. What was the last letter he wrote?

Problem 6.25 Neal arrived late at his friend's birthday party and everyone was playing a game. He didn't hear the rules so he was trying to figure out what the game was about. It was weird, they were counting starting from one, but sometimes they would clap. When someone didn't clap when they were supposed to the game was over and they started again: "one", "two", *clap*. "four", "five", *clap*, "seven", "eight", *clap*, "ten", "eleven", *clap*, *clap*, "fourteen", *clap*, ...

(a) What do you think are the rules to the game?

(b) After listening his friends play a couple of rounds Neal thought he finally got how the game was played, so he joined. Right before his turn someone said "twenty". What should Neil say/do in his turn?

(c) At some point during the game Neil hears them clapping ten times in a row. What was the next number they said out loud?

Problem 6.26 Find the missing numbers in the following pattern:

1	2	3	0
4	5	6	?
7	8	9	6
−2	−1	?	−3

Problem 6.27 Find the missing numbers in the following pattern:

1	3	4	12
2	5	?	70
6	7	2	84
12	?	56	70560

Problem 6.28 Nick and two of his friends went to an amusement park and are planning on getting on a roller coaster. There are a lot of people in line and the roller coaster has 8 rows of 4 seats each, labeled A, B, C and D, that are filled in order. The first four people in line go on the first row, the next four people go on the second row, and so on. In each row, the people fill in the seats in the same order: A, B, C and D.

(a) If there are 43 people in front of Nick and his friends are right after him in line, are they going to be sitting on the same row in the rollercoaster?

(b) If two people cut in line ahead of Nick. What row and seat is Nick going to be sitting at?

Problem 6.29 Kori is at it again. This time she found some triangle stickers and decided to make some figures with them following a pattern. The third and fourth figures she made look like this:

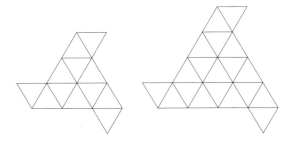

(a) What were her first two figures?

(b) Describe the pattern that Kori is following. How many triangle stickers does she use for each figure? Make sure to describe the pattern so that another person can read it and use it to think about the figure.

(c) How many triangle stickers did she use for her seventh figure?

Problem 6.30 Label the balls in the pyramid with the numbers of Pascal's triangle. Note that you may only have to work hard for half of them and the rest of them will be easy to find. Can you see why?

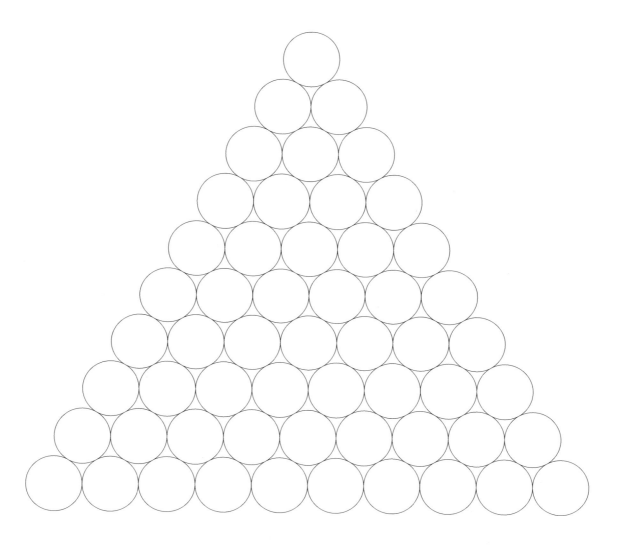

Problem 6.31 In the diagram of Pascal's triangle you filled out above, find

   (a) The counting numbers.

   (b) The sequence of triangular numbers.

Problem 6.32 Look again at the diagram of Pascal's triangle.

   (a) Find two balls that are touching each other that add up to $2^2 = 4$.

   (b) Find two balls that are touching each other that add up to $3^2 = 9$.

   (c) Find two balls that are touching each other that add up to $4^2 = 16$.

(d) Use the balls you identified to figure out a pattern to find the square numbers in Pascal's triangle.

Problem 6.33 We will use once more the diagram you figured out for Pascal's triangle. For each of the following questions we will add some of the numbers in the triangle. It might be great if you color the number we use in each question with a different color so you can identify a pattern. Start by coloring the top ball with one color and the first ball of the second layer with a different color.

(a) Add the numbers in the first ball of the third layer and the second ball of the second layer (use another color to paint these two balls).

(b) Add the numbers in the first ball of the fourth layer, and the second ball of the third layer (use another color to paint these two balls).

(c) Add the numbers in the first ball of the fifth layer, the second ball of the fourth layer and the third ball of the third layer (use another color to paint these three balls).

(d) Add the numbers in the first ball of the sixth layer, the second ball of the fifth layer and the third ball of the fourth layer (use another color to paint this three balls).

(e) Following this pattern, what numbers should you add up next? What is their sum?

(f) Continue following the pattern and make a list of the numbers you add up in each step. Keep that list and come back to this problem after you go through the problems in the next chapter. You'll see then why we wanted to follow this pattern.

Problem 6.34 In the following diagrams label the triangles that are pointing upwards with the numbers of Pascal's triangle, starting with a 1 on the uppermost triangle. Use a color to fill in the triangles that have odd numbers (odd numbers are those that aren't divisible by 2). See if you can identify a pattern before you color the last diagram.

(a)

(b)

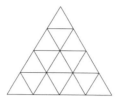

(c)

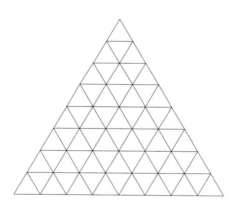

(d)

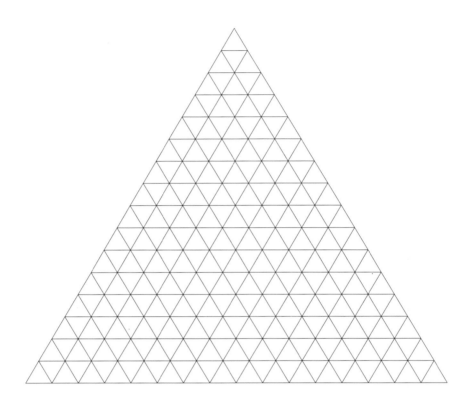

Problem 6.35 In the following diagrams label the triangles that are pointing upwards with the numbers of Pascal's triangle, starting with a 1 on the uppermost triangle. Use a color to fill in the triangles that have numbers that are not divisible by 3. See if you can identify a pattern before you color the last diagram.

(a)

(b)

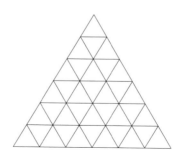

(c)

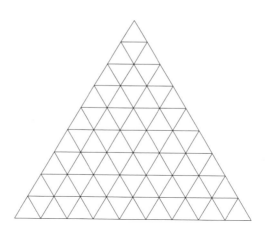

(d)

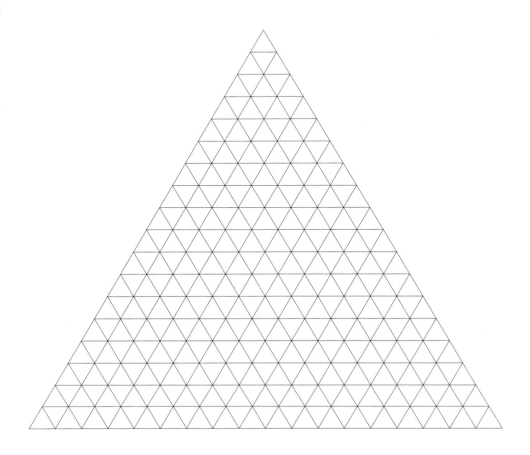

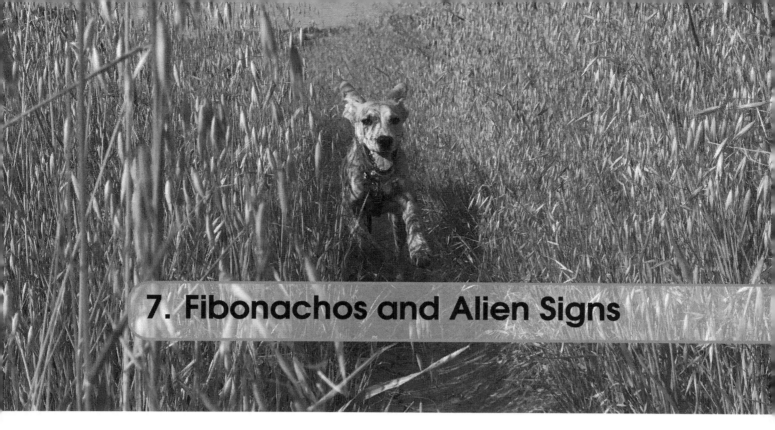

7. Fibonachos and Alien Signs

Remember Old Mr. Farmer Guy? Well, one day he woke up in his farm ready to harvest the corn in his field only to find that someone (aliens!?) had made a pattern in the middle of his field. It sort of looked like a giant letter S surrounded by a circle right in the middle of the field

He decided to harvest his crops anyway and started passing with his tractor truck right through the middle of the field, splitting the big S in 4 parts, then he passed again, and again, always vertically splitting the S into more and more parts. He only passed a few times, but still wondered "How many parts would that S be split into after I pass with my tractor truck 50 times?".

In this chapter we continue recognizing patterns and sequences, learning about some of the more famous types of sequences, such as arithmetic sequences, geometric sequences, and the Fibonacci sequence. These sequences show up all the time in everyday life, possibly even when you're eating some tasty nachos!

The concepts introduced in this chapter directly correspond to Common Core Math Standards as shown in the following table.

3rd Grade	3.OA.3, 3.OA.9
4th Grade	4.OA.2, 4.OA.3, 4.OA.5
5th Grade	5.OA.3

In addition to the standards above, problems and concepts in this section will help strengthen understanding of the following domains.

3rd Grade	3.OA, 3.NBT
4th Grade	4.OA, 4.NBT
5th Grade	5.OA, 5.NBT

7.1   Example Questions

Example 7.1

Let's help out Old Mr. Farmer Guy figure out what would happen with the giant letter S he found if he were to pass 50 times through his field with his tractor truck. Remember, when he found the field it looked like this from above

Solution

After passing one time with his tractor, the crop looks now like this so the giant letter S was split into 4 parts. After a few more times he passes with his tractor truck it looks like this

So, after 2 passes the S is now split into 7 parts, and into 10, 13, and 16 parts after 3, 4 and 5, respectively. Have you noticed a pattern already? Let's put our findings together on a table, that always helps.

# of passes	0	1	2	3	4	5	$\cdots$
# of parts	1	4	7	10	13	16	$\cdots$

Note that every time he passes again, 3 new parts are added since exactly 3 parts he already had are split in two parts each. This means that after 50 passes, he would have added 3 parts each time to the 1 part he started with. So, in total he would have that the giant S splits into

$$1 + 3 \times 50 = 151$$

parts.

Example 7.2

Determine the 20th number in this sequence: 9, 13, 17, 21, 25, ...

Solution

One thing that may help us spot a pattern when we have a sequence of numbers is look at the difference we get with every pair of consecutive numbers. The difference of the consecutive numbers we can see so far are

$$13 - 9 = 4 \quad 17 - 13 = 4 \quad 21 - 17 = 4 \quad 25 - 21 = 4$$

In each case we got the same difference of 4. This tells us that every time we want to add a new number to the list we just add 4 to the last one we have so far. The 20th number will be 19 places after the 9. That means we have to add $4 \times 19 = 76$ to 9 to get the number in the 20th position. So, the number we are looking for is $9 + 76 = 85$.

> ### Remark
>
> The previous two examples are examples of *arithmetic sequences*. This kind of sequences start with some number and each of the next numbers can be found by adding the same number over and over again.
>
> If you know some term in an arithmetic sequence and the *common difference*, you can go back and forth by adding or subtracting the common difference.

Example 7.3

Find the sum of the first 15 numbers in the sequence 2, 5, 8, 11, 14, ...

Solution

We can see that the sequence is an arithmetic sequence since the difference between each pair of consecutive terms is always 3. We want to sum up to the 15th term, which is

$$2 + 14 \times 3 = 44,$$

so we want to find the long sum

$$2 + 5 + 8 + \cdots + 38 + 41 + 44$$

There is an easier way to add all this numbers than just adding them all one by one. Look at what happens when we add the 1st term with the 15th, the 2nd with the 14th, and so on:

2	+	5	+	$\cdots$	+	20	+	23	+	26	+	$\cdots$	+	41	+	44
44	+	41	+	$\cdots$	+	26	+	23	+	20	+	$\cdots$	+	5	+	2

46	+	46	+	$\cdots$	+	46	+	46	+	46	+	$\cdots$	+	46	+	46

Each time we get exactly the same sum! The sum of all this numbers together would be 46 times the number of terms we have, so $15 \times 46 = 690$. But, when we added the numbers like this, we had each summand twice, so we need only half of that sum. This way, the sum of the first 15 numbers in the sequence is $690 \div 2 = 345$.

Remark

This would work with any arithmetic sequence and for any number of terms. We could even say we just found a nice "formula" for doing this:

$$\text{Sum} = \frac{(\text{First term} + \text{Last term}) \times \text{Total number of terms}}{2}$$

Example 7.4

Take a look again at the previous problem. What is the average value of the first 15 numbers in the sequence? ♣

Solution 1

We know that the average of a bunch of numbers is just their sum divided by how many numbers we have. In this case we already know that their sum is 345 and we have 15 of them, so the average is $345 \div 15 = 23$

Solution 2

Since we know that all the numbers are in an arithmetic sequence, there's an easier way to find the average without having to add them all up and then divide: When we found the sum in the previous problem, we multiplied 46 by 15 and then divided by 2. To find the average we are going to divide that sum by 15, but we know that $15 \div 15 = 1$ so we would get the same result if we avoid the step where we multiply by 15 when finding the sum. So the average is just $46 \div 2 = 23$. Note that this is the number right in the middle of our sequence!

Remark

> Be careful! This way of finding the average works only for arithmetic sequences! Also, note that the average in this example was just the number in the middle because we had an odd number of terms. If we had an even number of terms we would need to find the average of the two numbers in the middle of the sequence.

Example 7.5

Calculate $10 \times 10 - 9 \times 9 + 8 \times 8 - 7 \times 7 + \cdots + 2 \times 2 - 1 \times 1$ ♣

Solution

Note that all the terms in here are square numbers. We already know that the first 10 odd numbers add up to the 10th square number, 10×10, and the first 9 odd numbers add up to the 9th square number, 9×9. So, we have an easy way to subtract those two square numbers:

$$10 \times 10 - 9 \times 9 = (1 + 3 + \cdots + 17 + 19) - (1 + 3 + \cdots + 16 + 17) = 19.$$

If we do this again with each of the next differences we have in here, we have that it is the same as the sum of every other odd number from 3 to 19, so 3, 7, 11, 15 and 19. Note that this numbers form an arithmetic sequence, we have 5 of them and the one in the middle is 11. The sum of an odd number of consecutive terms of an arithmetic sequence is just the middle term times the number of terms, so this time the sum is $11 \times 5 = 55$

Example 7.6

Find the first term of the following arithmetic sequences.
  (a) An arithmetic sequence with 9 terms where the sum of all of the numbers is 405 and the difference between consecutive numbers is 3.
  (b) An arithmetic sequence with 6 terms where the sum of all of the numbers is 306 and the difference between consecutive numbers is 4. ♣

Solution to Part (a)

Since the sum of the numbers is 405, the term in the middle is $405 \div 9 = 45$. The term in the middle is the fifth term in the list, so we need to go back 4 numbers counting by 3 to get to the first number: 45, 42, 39, 36, 33.

Solution to Part (b)

We can do something similar as in the previous one, but this time we need to be careful about the "term in the middle" since we have en even number of terms. If we divide 306 by 6 we get the number 51. This would not be one of the numbers in our sequence, but it will be right in between the two terms in the middle, the third and fourth terms. Since we know that the difference between any two consecutive numbers in our list is 4, that means that 51 will be 2 more than the third term, so the third term in our list is 49. To find the first one we just need go back 2 numbers counting by 4: 49, 45 and 41.

Example 7.7

Robin started collecting pogs. She started with just 1 pog, and for the next weeks she plans to double the number of pogs she buys per week.
(a) How many pogs will Robin buy each of the next 5 weeks?
(b) How many pogs will Robin have gathered in the next 8 weeks?

Solution to Part (a)

We know she buys 1 pog this week. As she plans to double the amount she buys each week, the amount of pogs she buys a certain week will be 2 times the amount she bought the previous week. It will be easier to keep track of the amount of pogs she buys each week if we make a table. We will say "this week" is week 0 and start counting from next week:

Week	0	1	2	3	4	5
# of pogs bought	1	2	4	8	16	32

So, she will buy 2, 4, 8, 16 and 32 pogs each of the next 5 weeks.

Solution 1 to Part (b)

This time we want to make sure we look at the total number she has, not just the amount she bought. Let's use the same table as before, but we will add a row for the total number of pogs she has each week.

Week	0	1	2	3	4	5	6	7	8
# of pogs bought	1	2	4	8	16	32	64	128	256
Total # of pogs	1	3	7	15	31	63	127	255	511

Solution 2 to Part (b)

Did you notice that the total number of pogs she bought each week was exactly one less than the number of pogs she would buy the following week? If you spot this pattern early it is easy to figure out the total number without adding all the numbers. We would just need to figure out how many she would need to buy in week 9, so

$$\underbrace{2 \times 2 \times \cdots \times 2}_{9 \text{ times}} = 2^9 = 512$$

and subtract one from that number, so she would have $512 - 1 = 511$ pogs in total in 8 weeks.

Remark

This is an example of a *geometric sequence*. This kind of sequences start with a number and the following numbers are found by multiplying the previous number in the list by a fixed number, usually called the *common ratio*. In our example we started with 1 and the ratio we had was 2.

To go forth in the sequence one just needs to multiply the current number by the common ratio. To go back in the sequence, we can divide the current term by the common ratio.

Example 7.8

Figure out the first term of the following geometric sequences, assuming the terms in the sequences are all positive numbers.

(a) A geometric sequence where the third term is 45 and the fourth term is 135.

(b) A geometric sequence where the second term is 12 and the fourth term is 192.

Solution to Part (a)

Since we know two consecutive terms of the geometric sequence, we can figure out what is the common ratio of the numbers in it. We just need to divide the fourth number by the third number, so the common ratio is $135 \div 45 = 3$. This means that the second number in our sequence will be $45 \div 3 = 15$, and so the first number is $15 \div 3 = 5$.

Solution to Part (b)

This time we do not have consecutive terms, since we have the second and the fourth, but not the third. We can still figure out the problem, though. If we divide the fourth term by the second term we will find the square of the common ratio of our numbers: $192 \div 12 = 16$. This means that the common ratio of our numbers is 4, since $4 \times 4 = 16$. (The common ratio could also be -4, but since the sequence contains positive numbers, the common ratio is positive.) Now that we have the common ratio we can complete our list and find the first number. Just to make sure that we got the right number, let's start dividing the fourth term by 4 to find the third, second and first terms (by doing this we should get that the second term is 12, as we already know).

$$192 \div 4 = 48, \quad 48 \div 4 = 12, \quad 12 \div 4 = 3$$

Thus, the first term in our sequence is 3.

Example 7.9

Dr. Ezra was fascinated with some magical toads that had the ability to grow multiple heads. The only thing they needed to do was to jump over a toad of a different color from theirs and they would acquire as many heads as that toad. A blue and a yellow toad, each with one head, started taking turns jumping over each other. If the blue toad jumped first, how many heads will the yellow toad have after each of them jumped three times?

Solution

Let's help Dr. Ezra keep track of the number of heads that each toad has after jumping over each other. In the following diagrams we will represent the blue toad with a small circle and the yellow toad with a big circle. The first time the blue toad jumps he gets 1 head, for a total of $1 + 1 = 2$:

The first time the yellow toad jumps, he gets 2 heads from the blue toad, for a total of $1 + 2 = 3$

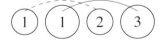

Following this pattern, the yellow toad would have 21 heads after each of them jumps three times:

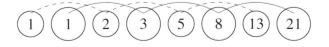

Remark

The number of heads the toads have are given by the *Fibonacci sequence*. It is a sequence of numbers that starts with 1, 1, and after that each of the numbers is the sum of the previous two numbers:

$$1, \quad 1, \quad 2, \quad 3, \quad 5, \quad 8, \quad 13, \quad 21, \quad \ldots$$

Example 7.10

Donnie loves nachos and Fibonacci numbers, so she opened up a *Fibonacho* restaurant. She had only one rule: only serve orders that a guest could eat in a Fibonacci style. So, eat one nacho, then one more, then two nachos, then three, five and so on.

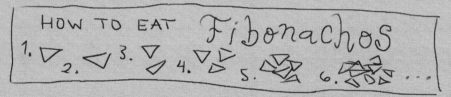

(a) A guest wants to order a plate that he can finish after eating nachos only 3 times. How many nachos will his plate have?

(b) Donnie wants to write the first few possible orders on a board to display at the store. She noticed that normally a customer would not be willing to eat more than 13 nachos at once. What numbers should Debbie list on her board?

(c) Roy held the record for the most nachos eaten at once. He ate 55 nachos in one single bite, right after eating 34 nachos. How many nachos did he eat in total that day?

Solution to Part (a)

Since this customer only wants to have three bites, we need only the first three Fibonacci numbers. They are are 1, 1 and 2, so the order should have $1 + 1 + 2 = 4$ nachos in total.

Solution to Part (b)

Since the usual customer does not eat more than 13 nachos at once, we will just need to list all Fibonacci numbers up to the one that is equal to 13 (that is the seventh Fibonacci number). Let's take a look at the total number of nachos in orders that have up to 7 bites available:

# of bites in order	1	2	3	4	5	6	7
# of nachos per bite	1	1	2	3	5	8	13
# of nachos in order	1	2	4	7	12	20	33

So, Donnie should write in her board that the most common orders include 1, 2, 4, 7, 12, 20 or 33 nachos, each.

Solution to Part (c)

Look closely at the table we set up in the previous part. The total number of nachos in an order is the same as 1 less than the Fibonacci number two places ahead. So, since we know that Roy ate 34 nachos, and then 55 nachos, his order should have 1 less nacho than the second Fibonacci number after 55. The next two Fibonacci numbers following 34 and 55 are $34 + 55 = 89$ and $55 + 89 = 144$. So his order had a total of $144 - 1 = 143$ nachos.

7.2 Quick Response Questions

Problem 7.1 Find the sum of all the counting numbers from 1 to 10.

Problem 7.2 If we start with 3 and count every fifth number, what will be the tenth number we count?

Problem 7.3 If we start with 3 and we keep multiplying by 11, what would be the fourth number in our list?

Problem 7.4 What is the next number in the sequence?

(a) 5, 8, 11, 14, ...

(b) 21, 26, 31, 36, ...

(c) 125, 150, 175, 200, ...

Problem 7.5 What is the sum of the first 10 even counting numbers?

Problem 7.6 What is the next term in the sequence?

(a) 1, 10, 100, 1000, 10000, ...

(b) 1, 4, 16, 64, 256, ...

(c) 1, 11, 121, 1331, 14641, ...

Problem 7.7 Find the next number in each of the following sequences:

(a) 3, 7, 10, 17, 27, ...

(b) 5, 2, 7, 9, 16, ...

(c) 2, 3, 5, 8, 13, ...

Problem 7.8 The average of three consecutive counting numbers is 55. What is the smallest of the three numbers?

Problem 7.9 What is $13 \times 13 - 12 \times 12$ equal to?

Problem 7.10 What is the sum of the first 8 Fibonacci numbers?

7.3 Practice

Problem 7.11 Sammy Sandstone was taking some pictures on a field trip to the savanna. At first he could only spot one lion, so he took his first picture. Two more lions approached and he took a second picture, capturing all three lions at once. It seems that more lions were arriving, two at a time, and Sammy took a picture every time there were new lions on the field.

(a) How many lions did Sammy have in his seventh picture?

(b) How many lions did Sammy have in all seven of his pictures?

Problem 7.12 Jen, my dog, used to like eating my socks. From January to June she ate 54 of my socks. I know that each month she ate 2 more than the previous month. How many socks did she eat in June?

Problem 7.13 3 letters arrived in the mail on Monday. 5 letters arrived in the mail on Tuesday, and 7 arrived on Wednesday. If this pattern continues until Sunday. How many pieces of mail will have accumulated by Sunday?

Problem 7.14 Corry went to the forest every day for a week to collect rocks. The first day he collected 5 rocks. Each of the following days he collected 2 more rocks than the previous day. What is the average of the number of rocks he collected each day?

Problem 7.15 Five successive counting numbers have a sum of 70, what are the numbers?

Problem 7.16 The aliens did it again! This time Old Mr. Farmer Guy came out to his field and found a big number 8 in the middle of it surrounded by a circle.

(a) Fill in the table: Into how many parts is the number 8 separated if Mr. Farmer Guy passes 2, 3, 4 and 5 with his tractor truck vertically close to the middle of the number 8?

# of passes	1	2	3	4	5
# of parts	2				

(b) Into how many parts is the 8 separated if Mr. Farmer Guy passes with his tractor truck 40 times?

(c) If the number 8 is now divided into 152 parts, how many times did Mr. Farmer Guy pass with his tractor truck?

Problem 7.17 Suppose you have a secret. On Sunday you tell it to a friend. On Monday your friend tells your secret to 2 other friends. On Tuesday, each of your friends who heard your secret on Monday tell it to 2 other friends. If this procedure continues, how many people in all will have been told your secret by the end of the following Sunday? Assume no one is told your secret more than once.

Problem 7.18 Henry used his two favorite numbers to write a sequence like the Fibonacci sequence in a piece of paper (that is, he added the two numbers to get a third number, then he added the second number and this new third number to get the fourth, and so on). His best friend, Eustace, has always been curious about Henry's favorite numbers so he took the piece of paper with the list of numbers to find out what they are, only to discover that Henry's dog had eaten the first few numbers off the list. Eustace could see that the 5th number on the list was 109 and the 6th number on the list was 167. Can you help him figure out Henry's favorite numbers?

Problem 7.19 Randia was visiting a friend at his house but he was late getting back from school. While Randia waited for her friend she decided to add up the numbers of the 7 houses in her friend's street. The house numbers are all consecutive and she got a sum of 2205. What are the numbers of the houses?

Problem 7.20 The sum of 6 consecutive even numbers is 234, what is the smallest of the numbers?

Problem 7.21 A list contains 7 consecutive even numbers. If their sum is 294, what is the second number on the list?

Problem 7.22 The following is an arithmetic sequence with some numbers missing. What are the missing numbers?

$$14, \underline{\hspace{1cm}}, \underline{\hspace{1cm}}, 26, \ldots$$

Problem 7.23 The fourth and sixth terms of a geometric sequence are 135 and 1215. Suppose the terms of this sequence are all positive, what is the fifth term of the sequence?

Problem 7.24 Find the sum of all the counting between the indicated limits:

(a) 3 through 12

(b) 15 through 55

(c) 5 through 90

Problem 7.25 The third term in a geometric sequence is 20, and the fourth term is 40. What is the sixth term?

Problem 7.26 The following is a geometric sequence with some numbers missing. What are the missing numbers?

$$12, \underline{\hspace{1cm}}, \underline{\hspace{1cm}}, 324, \ldots$$

Problem 7.27 Ashley wants to find the average temperature in a week. She checked the thermometer outside her window the first thing every morning, and recorded the following table:

Monday	Tuesday	Wednesday	Thursday	Friday	Saturday	Sunday
66°	67°	70°	69°	71°	68°	72°

What is the average temperature in the week?

Problem 7.28 Jing and Xing went to the Fibonachos restaurant and ordered a plate of 54 nachos for them to eat together. Jing ate one nacho and then they took turns eating.

(a) How many nachos did Jing eat?

(b) How many nachos did Xing eat?

(c) They wondered if they could have eaten the nachos in a different order (so not alternating) and still each of them eat the same number of nachos as they did before. Is this possible? (Remember: since they ordered a single plate, each time either of them eats nachos, they have to eat as many as the next Fibonacci number on the list).

Problem 7.29

(a) What is $51 \times 51 - 50 \times 50$ equal to?

(b) What is $49 \times 49 - 48 \times 48$ equal to?

(c) Calculate $51 \times 51 - 50 \times 50 + 49 \times 49 - 48 \times 48 + \cdots + 11 \times 11 - 10 \times 10$

Problem 7.30 Kathleen loves visiting her grandma because she has one of those grandfather's clocks that has a pendulum and chimes at the start of every hour. The clock chimes as many times as the hour. If Kathleen was at her grandma's house from 4:30 pm to 10:30 pm, how many chimes did she hear?

Problem 7.31 Darren was counting by threes starting at 5. His brother Phillip was adding up all the numbers. If If Darren counted 12 numbers, what sum did Phillip get?

Problem 7.32 The sum of ten consecutive even numbers is 370. What is the biggest number that was added?

Problem 7.33 Manny was practicing to be a magician. He got an egg at the magician's store but he dropped it through a flight of stairs. The magic egg didn't break, but it split into three eggs and every time the eggs reached the next step each of them would split into three eggs. How many eggs could Manny see at each of the 4 steps the egg fell through?

Problem 7.34 What is the next term in each of the following sequences?

(a) 20, 60, 180, 540, …

(b) 15, 60, 240, 960, …

(c) 34, 170, 850, 4250, …

Problem 7.35 Fill in the blank with the missing term in the sequence.

(a) 4, 12, 36, _____, 324, …

(b) _____, 308, 2156, 15092, …

(c) 12, 132, _____, 15972, 175692, ...

8. Chickens and Rabbits

There are some chickens and some rabbits on a farm. You know there are 65 heads and 180 feet in total among the chickens and rabbits. Without counting, can you figure out how many of the heads come from chickens and how many come from rabbits?

The "chicken and rabbit" method, based on wisdom from ancient China, allows you to quickly solve the problem above without mistakes. This creative method depends on your imagination (also, basic arithmetic operations of addition, subtraction, multiplication, and division) and can be used to solve many different problems. Take a look at the rest of this chapter to see the method in action!

The concepts introduced in this chapter directly correspond to Common Core Math Standards as shown in the following table.

3rd Grade	3.OA.3, 3.OA.8, 3.MD.3
4th Grade	4.OA.1, 4.OA.2, 4.OA.3

In addition to the standards above, problems and concepts in this section will help strengthen understanding of the following domains.

3rd Grade	3.OA, 3.NBT, 3.MD
4th Grade	4.OA, 4.NBT, 4.MD
5th Grade	5.OA, 5.NBT, 5.MD

8.1 Example Questions

Example 8.1

There are some chickens and some rabbits on a farm. Suppose there are 65 heads and 180 feet in total among the chickens and rabbits, how many chickens are there? How many rabbits?

Solution 1

First of all, each animal has one head. A chicken has two feet, and a rabbit has four feet. There are 65 heads, so there are 65 animals.

Now assume all 65 animals are chickens, there should be total 130 feet. But there are $180 - 130 = 50$ additional feet. These 50 feet came from the rabbits which were incorrectly assumed to be chickens. Each rabbit has 2 more feet than the incorrectly assumed chicken, so we need to count in two additional feet for each rabbit. Now $50 \div 2 = 25$, which is the number of rabbits. The rest is the number of chickens, which is $65 - 25 = 40$.

In conclusion, there are 25 rabbits, and 40 chickens.

Solution 2

Step 1. We may use a little bit more imagination this time. Imagine all the animals are trained with special skills for a circus performance. When the trainer blows a whistle, each chicken stands with one foot, and each rabbit stands up with two hind legs. All the animals lift half the number of feet up to the air. On the ground now, there should be $180 \div 2 = 90$ feet.

Step 2. Now each chicken has exactly one foot on the ground. However many chickens we have, we should count exactly same number of feet. But each rabbit would have two feet on the ground. Therefore, the extra number of feet on the ground 90, in comparison to the number of heads given, which is 65, would be the number of rabbits: $90 - 65 = 25$.

Step 3. We now know that there are 25 rabbits, so the number of chicken is $65 - 25 = 40$.

In conclusion, there are 25 rabbits, and 40 chickens.

Example 8.2

In a math competition, there are 30 questions. Each correct answer earns 5 points. One point is taken away for each incorrectly answered or unanswered question. Jenny received 114 points. How many questions did she answer correctly?

Solution

If Jenny answered all questions correctly, she would have gotten a perfect score of

$$30 \times 5 = 150.$$

However, her resulting score was 114. For each incorrect or blank answer, Jenny loses the 5 points that the she would have gotten for the question and an additional point for

192 Chapter 8. Chickens and Rabbits

the incorrect answer. Thus, her score decreases by a total of

$$5 + 1 = 6$$

points for each question missed. The total decrease in the score was

$$150 - 114 = 36$$

points, and thus the number of incorrectly answered or unanswered question was

$$36 \div 6 = 6.$$

Therefore Jenny answered

$$30 - 6 = 24$$

questions correctly.

Example 8.3

The price of a pack of colored pencils is $17 and the price of a pack of regular pencils is $12. The math teacher bought 24 packs of pencils for a total of $373. How many packs of each type did he buy?

Solution

If all 24 packs were colored pencils, then the teacher would have spent

$$17 \times 24 = 408$$

dollars, which is

$$408 - 373 = 35$$

more dollars than the actual amount. Each pack of regular pencils is

$$17 - 12 = 5$$

dollars less expensive than colored pencils, and therefore the math teacher bought

$$35 \div 5 = 7$$

packs of regular pencils. The remaining

$$24 - 7 = 17$$

packs are colored pencils.

Remark

> Did you notice in this problem that we pretended the total money spent to be the "total number of feet" and the cost of each kind of pencil to be the "number of feet" of each colored pencil?

Example 8.4

Tim and Will have to type up a report that contains 600 words. Tim can type 50 words per minute, but Will can only type 40 words per minute. Will starts typing the report, and after some time Tim takes over to speed up the process. Together it takes a total of 13 minutes to type the report. How many minutes does Tim type? What about Will?

Solution

If Will had typed by himself for 13 minutes, he would have typed

$$13 \times 40 = 520$$

words in total. That is, we would be still missing

$$600 - 520 = 80$$

words. Since Tim can type $50 - 40 = 10$ more words than Will every minute, if Tim types for

$$80 \div 10 = 8$$

minutes, he would make up for the missing words. Hence Tim typed for 8 minutes and Will typed for $13 - 8 = 5$ minutes.

Remark

> This time the total number of minutes spent was the "total number of heads", the number of words in the report was the "total number of feet" and the number of words per minute each of them could type was the "number of feet" that each of them had.

Example 8.5

Every weekend Judy volunteers at the local soup kitchen, where she helps wash dishes. Before she starts, Judy counts 80 dishes that she needs to wash. Judy can wash 5 dishes per minute, and begins washing the dishes. Before she has finished, her friend Suzy joins to help her out, who can also wash 5 dishes per minute. When they finish, Judy notes she has washed dishes for a total of 12 minutes. How many minutes did Suzy help?

Solution

After 12 minutes, Judy alone would have only washed

$$5 \times 12 = 60$$

dishes, that is, she would still have

$$80 - 60 = 20$$

dishes to wash. Since Suzy can wash 5 dishes per minute, if she helps for

$$20 \div 5 = 4$$

minutes they would be done in 12 minutes.

Example 8.6

To make the football team, George must run 20 laps within 30 minutes. If George runs, it takes 1 minute for him to complete a lap. If he walks, it takes him 3 minutes to complete a lap. George likes football, but does not like running, so he decides to figure out how much he can walk so that it takes him exactly 30 minutes to finish the 20 laps. If he finishes in exactly 30 minutes, how many laps did George walk? How many did he run?

Solution

If George only walks, if would take him

$$20 \times 3 = 60$$

minutes to complete the 20 laps. That is

$$60 - 30 = 30$$

more minutes than he should spend on the course. When he walks, it takes him $3 - 1 = 2$ more minutes to complete a lap than when he runs, so if he ran for

$$30 \div 2 = 15$$

minutes, he could make up for that extra time. Thus, if he walks for $30 - 15 = 15$ minutes and runs for 15 minutes, he should complete the 20 laps. This way, he walks for $15 \div 3 = 5$ laps and runs for $15 \div 1 = 15$ laps.

Example 8.7

There are some chickens and some rabbits on a farm. Suppose there are 70 heads and there are 40 more rabbit feet than chicken feet on the farm. How many chickens and how many rabbits are there?

Solution

If all 70 animals are rabbits, there are

$$70 \times 4 = 280$$

rabbit feet and 0 chicken feet, so there are 280 more rabbit feet than chicken feet. If we replace a rabbit with a chicken, we remove 4 rabbit feet and add 2 chicken feet, so the difference between rabbit feet and chicken feet decreases by

$$4 + 2 = 6.$$

Since we know there are 40 more rabbit feet than chicken feet, which is

$$280 - 40 = 240,$$

smaller than if all the animals were rabbits, we must replace

$$240 \div 6 = 40$$

rabbits with chickens. Hence there are

$$70 - 40 = 30$$

rabbits and 40 chickens on the farm.

Example 8.8

150 monks eat 150 steamed buns. If each senior monk eats 2 steamed buns, and 2 junior monks eat 1 steamed bun, how many senior monks and junior monks are there?

Solution 1

Since two junior monks eat 1 steamed bun, we can split all steamed buns in half and pretend that each junior monk eats 1 half steamed bun and each senior monk eats 4 half steamed buns. This way 150 monks eat a total of $2 \times 150 = 300$ half steamed buns. If all 150 monks were senior, they would eat a total of

$$150 \times 4 = 600$$

half steamed buns, but that is

$$600 - 300 = 300$$

more half buns than there are available. We know that a senior monk eats $4 - 1 = 3$ more half steamed buns than a junior monk, so if we swap

$$300 \div 3 = 100$$

senior monks with junior monks, the monks will eat the correct number of half steamed buns. Therefore there are 100 junior monks and $150 - 100 = 50$ senior monks.

Solution 2

Grouping one senior monk with two junior monks, we have a group of 3 monks that eats

$$2 + 1 = 3$$

steamed buns in total. Since there are 150 steamed buns in total, there must be

$$150 \div 3 = 50$$

such groups. In each group, there is only one senior monk, so there are

$$50 \times 1 = 50$$

senior monks in total. Similarly, in each group, there are 2 junior monks, so there are

$$50 \times 2 = 100$$

junior monks in total.

Example 8.9

Suppose there are chickens, rabbits, and sheep on a farm. There are 90 heads in total and 288 feet. If there are the same number of rabbits and sheep, how many chickens, rabbits, and sheep are on the farm?

Solution

Since rabbits and sheep have 4 feet each, we can first assume there were no chicken, then there would be

$$90 \times 4 = 360$$

feet. However, there are in fact only 288 feet, so

$$360 - 288 = 72$$

feet are missing. This is because some of the animals are chicken. Each chicken reduces the number of feet by

$$4 - 2 = 2,$$

therefore there are

$$72 \div 2 = 36$$

chickens. Hence there are

$$90 - 36 = 54$$

animals with 4 feet each. Since there are the same number of rabbits and sheep, the number for each kind is

$$54 \div 2 = 27.$$

Thus there are 36 chickens, 27 rabbits, and 27 sheep on the farm.

Example 8.10

Ron is classifying bugs for a biology project. He has 17 bugs in total. Some are spiders, with 8 legs. Some are houseflies, with 6 legs and 1 pair of wings. The remaining are dragonflies, with 6 legs and 2 pairs of wings. If there are 114 legs and 18 pairs of wings, how many of each type of bug does Ron have?

Solution

If we first assume all 17 bugs are houseflies, then there will be

$$17 \times 6 = 102$$

legs. There are in fact 114 legs, which is

$$114 - 102 = 12$$

more than if we all bugs were houseflies. Each dragonfly has the same number of legs as a housefly but each spider has

$$8 - 6 = 2$$

more legs, so there must be

$$12 \div 2 = 6$$

spiders in total. The number of houseflies and dragonflies is thus

$$17 - 6 = 11.$$

Again, assume these 11 bugs are all houseflies. There will be 11 pairs of wings, which is

$$18 - 11 = 7$$

less than the actual amount. Each dragonfly has an extra pair of wings, so there must be 7 dragonflies. Hence the number of houseflies is

$$11 - 7 = 4.$$

To summarize there are 6 spiders, 4 houseflies, and 7 dragonflies.

8.2 Quick Response Questions

Problem 8.1 Michelle counts her chickens and rabbits, and there are 16 heads and 46 feet. How many of each type are there?

Problem 8.2 In a farm there are goats and ducks. The total number of heads is 100, and the total number of legs is 296. How many animals of each type are there?

Problem 8.3 Seventy vehicles (cars and motorcycles) are parked in a parking lot. Totally there are 190 wheels. Given that a car has 4 wheels and a motorcycle has 2 wheels, how many cars and motorcycles each are in the parking lot?

Problem 8.4 Teachers and students from the Areteem Summer Camp visited the museum. They bought a total of 100 tickets for 220 dollars. If each teacher ticket costs 4 dollars, and each student ticket costs 2 dollars, how many teachers and students were there respectively?

Problem 8.5 There are some chickens and some rabbits on a farm. Suppose there are 45 heads and 136 feet in total among the chickens and rabbits, how many of the animals are chickens? How many are rabbits?

Problem 8.6 Two trucks dump dirt of 500 cubic meters. Truck A carries 8 cubic meters per load. Truck B carries 4 cubic meters per load. The dirt is removed after 80 loads. How many loads are carried by truck A?

Problem 8.7 Each set of chess is played by 2 students, and each set of Chinese Checkers is played by 6 students. A total of 30 sets of chess and Chinese Checkers are played by exactly 140 students in a school event. How many sets of each game are there?

Problem 8.8 Use 400 matches to make triangles and pentagons. Totally 90 triangles and pentagons are made with no matches left over. How many of each shape are made?

Problem 8.9 Candace scored 40 points in her school's playoff basketball game. She made a combination of 2-point shots and 3-point shots during the game. If she made a total of 15 shots, how many 3-point shots did she make?

Problem 8.10 There are 48 tables in a restaurant. Small tables can seat 2 people, and big tables can seat 5 people. They can accommodate a maximum number of 156 people. How many small tables and how many big tables are there?

8.3 Practice

Problem 8.11 The counselor brought his 67 students to the lake to go rowing, 6 people for each big boat and 4 people for each small boat. They rented 14 boats to fit everyone with no empty seats. How many big and small boats each did they rent?

Problem 8.12 There are 25 questions in a math competition. 10 points are given to each correct answer, and 4 points are taken for each incorrect answer or unanswered question. Jeff received 180 points in the competition. How many questions did he answer correctly?

Problem 8.13 The company Green Pilots is organizing a company picnic at the beach. They want to save energy by carpooling to the beach site. They are able to fit 535 people into 95 vehicles that are either 5-seat sedans or 7-seat minivans. How many sedans and how many minivans do they need to use to take everyone to the picnic?

Problem 8.14 A large family of 30 people goes to a restaurant. They each order either pizza or salad. The pizza costs $11.00 and salad costs $6.00. In all the family spends $265.00. How many pizzas and how many salads did the family order?

Problem 8.15 Sasha takes a mathematics competition. There are a total of 15 problems. For each correct answer, competitors receive 5 points. For each wrong answer, they instead get 2 points taken away. Sasha has 33 points in total. How many problems did she answer correctly?

Problem 8.16 Aria has 20 coins that are nickels and dimes. The total value is $1.40. How many nickels and dimes does Aria have? (A nickel is 5 cents, and a dime is 10 cents.)

Problem 8.17 There are 150 birds and cats. There are 120 more bird legs than cat legs. How many birds and how many cats are there?

Problem 8.18 A group of 85 people rent 25 go-kart cars of two kinds at a racetrack. The first kind has a capacity of 3 people and costs $60 per car. The second has a capacity of 4 people and costs $75 per car. The 85 people exactly fill all vehicles. What is the total cost in renting the 25 go-kart cars?

Problem 8.19 There are many ducks and sheep in a farm. If we count the heads, there are a total of 60 heads. If we count the legs, there are 96 more legs from sheep than from ducks. How many ducks and how many sheep are there in the farm?

Problem 8.20 The Math Club collected donations from 52 people who live in either Landover or Salisbury. Each person from Landover contributed $10, and each person from Salisbury contributed $7. In total, $44 more were collected from Landover than from Salisbury. How many people contributed from each city?

Problem 8.21 In a farm the total number of chickens and rabbits is 80. If the number of chicken feet is 70 more than the number of rabbit feet, how many chickens and rabbits are there respectively?

Problem 8.22 A turtle has 4 legs and a crane has 2 legs. There are totally 110 heads of turtles and cranes, and there are 40 more crane legs than turtle legs. How many of each animal are there?

Problem 8.23 On a good day, Chris the Squirrel picks 30 hazelnuts. On a rainy day he only picks 15 hazelnuts. During a few consecutive days he picked a total of 225 hazelnuts with an average of 25 per day. How many days were rainy?

Problem 8.24 Morgan needed 85 sticks for a project at school. Each stick is either 4 inches long or 7 inches long and the total length of all the sticks combined is 430 inches. How many 4 inch sticks and 7 inch sticks are there?

Problem 8.25 A spider has 8 legs and no wings. A dragonfly has 6 legs and 2 pairs of wings. A cicada has 6 legs and one pair of wings. There are a total of 27 bugs of these types, with 180 legs and 28 pairs of wings in total. How many dragonflies are there?

Problem 8.26 Lily spent $465 to buy 67 color pencils for her art class, including red, green, and blue colors. The red pencils cost $3 each, the green ones cost $10 each, and the blue ones cost $7 each. Suppose she bought the same number of green and blue pencils. How many of each type of pencils did she buy?

Problem 8.27 A crab has 10 legs. A mantis has 6 legs and 1 pair of wings. A dragonfly has 6 legs and 2 pairs of wings. There are a total of 44 of the three types. There are 312 legs in total. There are 43 pairs of wings in total. How many of each kind are there?

Problem 8.28 Tony's mom took out $570 from the bank. There are $2, $5, and $10 bills and total of 120 bills. The number of $5 bills and $10 bills is the same. How many bills of each type are there?

Problem 8.29 60 mice eat 60 cakes. If each big mouse eats 4 cakes, and 4 baby mice eat 1 cake, how many big mice and baby mice are there?

Problem 8.30 A candy shop sells three flavors of candies: cherry, strawberry, and watermelon. The prices are $15/kg, $20/kg, and $25/kg, respectively. This morning the shop sold a total of 105 kg and received $2150. Given that the total sale of cherry and watermelon flavor candies combined was $1250, how many kilograms of watermelon flavor candies were sold?

Problem 8.31 John and Jane are holding a party at their home in the jungle. For the party, they need to collect 200 bananas. John can collect 35 bananas per hour, while Jane can collect 20 bananas per hour. Further, only one of them can collect bananas at a time, as someone must stay at home to watch their son Korak. John starts collecting the bananas and then Jane takes over to collect the rest. If a total of 7 hours is spent collecting bananas, how many hours are spent by Jane?

Problem 8.32 Jason and Perry work together to cut the grass at their school's soccer field. In total, the area of the field is 280 square yards. Jason can mow 60 square yards per hour and Perry can mow 40 square yards per hour. They start moving the lawn together, but Jason quits before the whole field is complete and Perry finishes the job working a total of 4 hours. How many hours did Jason work?

Problem 8.33 Fred and Ted are in charge of painting desks for Mrs. Larson's classroom, they want all the desks to be painted white. There are 30 desks in total, 19 of them are already white and the rest are gray. On the first day, only Ted painted, but he got confused, and started painting the already white desks gray. On the second day, only Fred painted, and by the end of the day all 30 desks (including the ones Ted had painted the wrong color) were white. Assume that Ted can paint 1 desk per hour and Fred can paint 2 desks per hour. If we know that altogether Ted and Fred painted for 16 hours, how many hours did Fred paint?

Problem 8.34 Maxime must drive 12 miles to work. Some of the trip is on the highway, where it takes him 1 minute to drive 1 mile. The rest of the trip is on city streets, where it takes 2 minutes to drive 1 mile. If his whole trip takes 20 minutes, how many miles does he drive on the highway? on city streets?

Problem 8.35 Alice completed 15 laps at the local track. Some of the laps she walked and the others she ran. Running, it takes Alice 2 minutes to complete a lap. Walking, it takes Alice 5 minutes to complete a lap. Disappointed in her time, the next day Alice completes the laps again, this time running every lap she walked the first time and walking every lap she ran the first time. This improves Alice's time by 15 minutes. How many laps did Alice run originally?

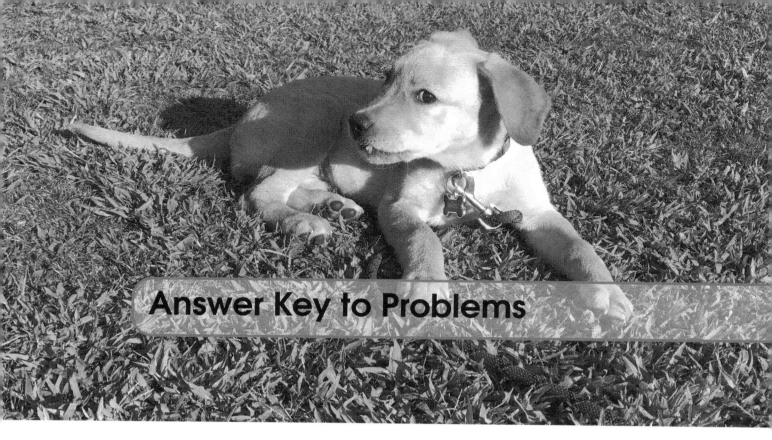

Answer Key to Problems

The answers to the quick response and practice problems in all the chapters are provided here. Only short answers are included. Full solutions for all problems can be found in the solutions manual, *Fun Math Problem Solving for Elementary School Solutions Manual*.

Chapter 1. It Matters Where It Is

Problem 1.1 (a) 235

  (b) 607

  (c) 192

Problem 1.2 Even

Problem 1.3 No

Problem 1.4 Even

Problem 1.5 Even

Problem 1.6 6

Problem 1.7 1

Problem 1.8 6

Problem 1.9 (a) 50; 5000

  (b) 30; 3000

  (c) 33; 3300

Problem 1.10 (a) 1209

  (b) 2214

  (c) 3065

Problem 1.11 (a) 2000 and 90

  (b) 3

  (c) 9000 and 8

  (d) 800

  (e) 4

Problem 1.12 (a) 8438

  (b) 7013

  (c) 7668

Problem 1.13 (a) 336

  (b) 2006

  (c) 5053

Problem 1.14 112, 212, 312, 412, 512, 612, 712, 812 and 912

Problem 1.15 890

Problem 1.16 284

Problem 1.17 23076

Problem 1.18 7624

Problem 1.19 Sample answer: 248

Problem 1.20 1348

Problem 1.21 165

Problem 1.22 25

Problem 1.23 99899

Problem 1.24 9867

Problem 1.25 (a) 9765210
  (b) 1025679
  (c) 9765201
  (d) 1025796

Problem 1.26 54963

Problem 1.27 2614 miles

Problem 1.28 1249

Problem 1.29 17

Problem 1.30 1498

Problem 1.31 3901

Problem 1.32 92

Problem 1.33 No

Problem 1.34 10, 11, 12, 13, and 14, or 11, 12, 13, 14, and 15

Problem 1.35 Sample answer: 201, 202 and 203

Chapter 2. Mathemagics

Problem 2.1 1 & 9; 2 & 8; 3 & 7; 4 & 6; 5 & 5; . . .

Problem 2.2 1 & 8; 2 & 7; 3 & 6; 4 & 5; 5 & 4; . . .

Problem 2.3 Sample answers: 64 & 35; 12 & 87

Problem 2.4 Sample answers: 67 & 33; 29 & 71

Problem 2.5 Sample answers: 564 & 436; 783 & 217

Problem 2.6 6 & 11; 8 & 9

Problem 2.7 (a) 228
  (b) 224
  (c) 234

Problem 2.8 (a) 473
  (b) 693
  (c) 363

Problem 2.9 (a) 324

(b) 361

(c) 289

Problem 2.10   (a) 4225

(b) 9025

(c) 5625

Problem 2.11   (a) 99

(b) 999

(c) 999

(d) 9999

(e) 9999

Problem 2.12   (a) 1099

(b) 10598

(c) 19999

Problem 2.13   (a) 8173

(b) 67892

(c) 35651

Problem 2.14   (a) 1157

(b) 24000

(c) 140970

Problem 2.15   (a) 5329

(b) 2000

(c) 400

Problem 2.16   (a) 1300

(b) 1000

(c) 6574

Problem 2.17   (a) 11025

 (b) 13225

 (c) 15625

 (d) 18225

 (e) 21025

 (f) 24025

 (g) 27225

 (h) 30625

 (i) 34225

Problem 2.18   (a) 3; 3

 (b) 18; 454

 (c) 18; 9; 450

Problem 2.19   (a) 6; 4; 64

 (b) 5; 3; 497

 (c) 6; 4; 6004

Problem 2.20   (a) 588

 (b) 6501

 (c) 9012

Problem 2.21   (a) 149

 (b) 393

 (c) 313

Problem 2.22   (a) 150

 (b) 62

 (c) 180

Problem 2.23   (a) 1540

 (b) 35255

Problem 2.24 4015

Problem 2.25   (a) 280

  (b) 36100

Problem 2.26   (a) 95

  (b) 87

Problem 2.27   (a) 289

  (b) 1225

  (c) 264

Problem 2.28   (a) 625

  (b) 550

  (c) 65

Problem 2.29   (a) 4221

  (b) 624

  (c) 7221

Problem 2.30   (a) 245021

  (b) 99209

  (c) 616221

Problem 2.31   (a) 48

  (b) 480

  (c) 240

Problem 2.32   (a) 2278

  (b) 7626

  (c) 2146

Problem 2.33   (a) 48856

(b) 164997

(c) 217848

Problem 2.34   (a) 158723

(b) 219538

(c) 2763486

Problem 2.35   (a) 6523

(b) 25874

(c) 13377

Chapter 3. Big and Small, Included

Problem 3.1 Aiden: 8; Brian: 4

Problem 3.2 90

Problem 3.3 No

Problem 3.4 Jerry: 15; Tom 45

Problem 3.5 40

Problem 3.6 $240

Problem 3.7 Female: 36; Male: 72

Problem 3.8 Aiden: 46; Brandon: 23

Problem 3.9 Hobbit: 30 inches; Elf: 90 inches

Problem 3.10 Duck: 12; Geese: 4

Problem 3.11 Friend's cake: 12 candles; Twin's cakes: 4 candles, each

Problem 3.12 10 liters

Problem 3.13 Bananas: 30 pounds; Basket: 2 pounds

Problem 3.14 Adam: 10 ; Bob: 4

Problem 3.15 Sinbad: 235; Popeye: 470; Captain Hook: 940

Problem 3.16 Mr. Giant: 8 feet; Mr. Super Giant: 37 feet

Problem 3.17 Girls: 17; Boys: 31

Problem 3.18 Lucy: $7; Jadean: $28

Problem 3.19 $10

Problem 3.20 Pencil: 30¢; Pen: $1.80

Problem 3.21 $165

Problem 3.22 Suzi: 8; Lisa: 16

Problem 3.23 Classical: 11; Pop: 44

Problem 3.24 Bob: 12; Bob's dad: 37

Problem 3.25 Old Wendy: 9 pounds; Old McDonald: 27 pounds

Problem 3.26 Oranges: 375 pounds; Bananas: 1225 pounds

Problem 3.27 Girls: 30; Boys: 40

Problem 3.28 Peg Legged Samuel: 14; Gold Toothed Brendan: 43

Problem 3.29 Clark Kent: 750 mph; Neo: 210 mph

Problem 3.30 Bugs: 18; Tweety: 6

Problem 3.31 Mickey: 16 miles; Donald: 10 miles; Goofy 32 miles

Problem 3.32 870

Problem 3.33 Harry: 10 ounces; Hermione: 18 ounces

Problem 3.34 576 square feet

Problem 3.35 $8

Chapter 4. Come Together, Leave Apart

Problem 4.1 30 and 35

Problem 4.2 6 feet

Problem 4.3 21

Problem 4.4 Patrick: 6; Joe: 16

Problem 4.5 Tim: 11; Allison: 9

Problem 4.6 Mike: 3 and a half; Josh: 6 and a half

Problem 4.7 31

Problem 4.8 20

Problem 4.9 33

Problem 4.10 5 and 11

Problem 4.11 68 years old

Problem 4.12 Olive: 21; Bluto: 7

Problem 4.13 40 and 60

Problem 4.14 Kyle: 16; Lisa: 8

Problem 4.15 Literature: 80; History: 70; Math: 100

Problem 4.16 9

Problem 4.17 Terrence: 20; Terrance's son: 12

Problem 4.18 Chris: 125; Bob: 250; Adam: 500

Problem 4.19 30

Problem 4.20 3rd grade: 34; 4th grade: 45; 5th grade: 29

Problem 4.21 Sword: 46 drachmas; Spear: 37 drachmas

Problem 4.22 Quiz A: 67; Quiz B: 46; Quiz C: 35

Problem 4.23 Old Sealegs Austin: 42; Landlubber Matthew: 35; Cap'n Poopdeck: 53

Problem 4.24 50

Problem 4.25 66

Problem 4.26 65 seconds

Problem 4.27 6 feet

Problem 4.28 Cameron: $80; Mandy: $130; Taylor: $110

Problem 4.29 4 volleyballs

Problem 4.30 21

Problem 4.31 Thyme: 4; Rosemary: 20

Problem 4.32 40 pounds

Problem 4.33 1075 and 1225

Problem 4.34 Spoons:25, Forks:15; Knives: 30

Problem 4.35 Savory Shreds: 12; Classic Paté: 8; Prime Filets:18

Chapter 5. Counting Without Fingers

Problem 5.1 7

Problem 5.2 10

Problem 5.3 24

Problem 5.4 27

Problem 5.5 36

Problem 5.6 16

Problem 5.7 3

Problem 5.8 9

Problem 5.9 6

Problem 5.10 28

Problem 5.11 36

Problem 5.12 73

Problem 5.13 197

Problem 5.14 1344

Problem 5.15 45

Problem 5.16 504

Problem 5.17 10

Problem 5.18 96

Problem 5.19 32

Problem 5.20 3

Problem 5.21 4

Problem 5.22 300

Problem 5.23 24

Problem 5.24 96

Problem 5.25 2625

Problem 5.26 4160000

Problem 5.27 100

Problem 5.28 5040

Problem 5.29 720

Problem 5.30 1440

Problem 5.31 19

Problem 5.32 19

Problem 5.33 10

Problem 5.34 60

Problem 5.35 112

Chapter 6. Figurate Numbers

Problem 6.1   (a) 36

  (b) 91

  (c) 42

Problem 6.2 1

Problem 6.3 2

Problem 6.4   (a) △

  (b) ★▷

  (c) △△△□□□

Problem 6.5 Saturday

Problem 6.6   (a) ◣

  (b) ◣

Problem 6.7 April

Problem 6.8 Monday

Problem 6.9 Friday

Problem 6.10 Thursday

Problem 6.11   (a) 4

  (b) 55555

  (c) 1001

Problem 6.12   (a) 1

  (b) 3

  (c) 6

  (d) 190

Problem 6.13   (a) Every time he makes a bigger square

  (b) 36

  (c) 13

Problem 6.14   (a) Add next odd number

(b) Square numbers

(c) 10000

(d) 2475

Problem 6.15   (a) ♡◇♡◇♡◇♡◇♡◇

(b) ⋆★★★★★⋆

(c) ←⇐⇐

Problem 6.16 3

Problem 6.17   (a) ⋆∪◁▷

(b) ↑↓↑↓↑

(c) ⋆|⋆||⋆|⋆

Problem 6.18   (a) 50

(c) 962

Problem 6.19 5 and 3

Problem 6.20 50

Problem 6.21   (a) △△△△△△△△△△△□□□□

(b) ↓↓↓↓↓↑↑↑↑↓↓↓↑↑↓↑↑↓↓↓↑↑↑↑↓↓↓↓

(c) ◁◁◁◁▷◁▷▷◁◁▷◁▷▷▷◁◁◁▷◁▷▷◁◁◁▷◁▷▷▷▷

Problem 6.22 1

Problem 6.23   (a) 94

(b) 115

Problem 6.24 S

Problem 6.25   (a) They must clap instead of saying the number if the number is a multiple of 3 or if the number has the digit 3 in it.

(b) *clap*

(c) "forty"

Problem 6.26 3 and 0

Problem 6.27 7 and 105

Problem 6.28 (a) No

(b) 4B

Problem 6.29 (a)

(b) If she is making figure n, she uses $n \times n + 3$ triangle stickers

(c) 52

Problem 6.30 Remember we need to start by writing 1s on the top and on the beginning and end of each row, then to find any other number in the triangle we need to add up the two numbers on top of it.

Problem 6.31 (a) The counting numbers are in the second diagonal (the one right next to the diagonal that has only 1s).

(b) All of the triangular numbers can be found in the third diagonal of the triangle, right below the counting numbers.

Problem 6.32 (a) 1 and 3 in the third and fourth layers

(b) 3 and 6 in the fourth and fifth layers

(c) 6 and 10 in the fifth and sixth layers

(d) The square numbers can be found by adding consecutive triangular numbers (the ones that we found were on the third diagonal of the triangle).

Problem 6.33 (a) 2

(b) 3

(c) 5

(d) 8

(e) $1 + 5 + 6 + 1 = 13$

(f) $1, 1, 2, 3, 5, 8, 13, 21, 34, 55, \dots$

Problem 6.34   (d)

Problem 6.35   (d)

Chapter 7. Fibonachos and Alien Signs

Problem 7.1 55

Problem 7.2 48

Problem 7.3 3993

Problem 7.4   (a) 17

  (b) 41

  (c) 225

Problem 7.5 110

Problem 7.6 (a) 100000

(b) 1024

(c) 161051

Problem 7.7 (a) 44

(b) 25

(c) 21

Problem 7.8 54

Problem 7.9 25

Problem 7.10 54

Problem 7.11 (a) 13

(b) 49

Problem 7.12 14

Problem 7.13 63

Problem 7.14 11

Problem 7.15 12, 13, 14, 15, 16

Problem 7.16 (a) 5, 8, 11, 14

(b) 119

(c) 51

Problem 7.17 255

Problem 7.18 44 and 7

Problem 7.19 312, 313, 314, 315, 316, 317 and 318

Problem 7.20 34

Problem 7.21 38

Problem 7.22 18 and 22

Problem 7.23 405

Problem 7.24   (a) 75
   (b) 1435
   (c) 4085

Problem 7.25 160

Problem 7.26 36 and 108

Problem 7.27 $69°$

Problem 7.28   (a) 21
   (b) 33
   (c) Yes

Problem 7.29   (a) 101
   (b) 97
   (c) 1281

Problem 7.30 45

Problem 7.31 258

Problem 7.32 46

Problem 7.33 3, 9, 27 and 81

Problem 7.34 (a) 1620

  (b) 3840

  (c) 21250

Problem 7.35 (a) 108

  (b) 44

  (c) 1452

Chapter 8. Chickens and Rabbits

Problem 8.1 Chickens: 9; Rabbits: 7

Problem 8.2 Goats: 48; Ducks: 52

Problem 8.3 Cars: 25; Motorcycles: 45

Problem 8.4 Teachers 10; Students: 90

Problem 8.5 Chickens: 22; Rabbits: 23

Problem 8.6 45

Problem 8.7 Chess: 10; Chinese Checkers: 20

Problem 8.8 Pentagons: 65; Triangles: 25

Problem 8.9 10

Problem 8.10 Small tables: 28; Big tables: 20

Problem 8.11 Big boats: 6; Small boats: 8

Problem 8.12 20

Problem 8.13 65 sedans, 30 minivans

Problem 8.14 Pizza: 17; Salads: 13

Problem 8.15 9

Problem 8.16 Nickels: 12; Dimes: 8

Problem 8.17 Birds: 120; Cats: 30

Problem 8.18 $1650

Problem 8.19 24 ducks; 36 sheep

Problem 8.20 Landover: 24; Salisbury: 28

Problem 8.21 Chickens: 65; Rabbits: 15

Problem 8.22 Turtles: 30; Cranes: 80

Problem 8.23 3

Problem 8.24 4 inch: 55; 7 inch: 30

Problem 8.25 10

Problem 8.26 Red: 19; Green: 24; Blue 24

Problem 8.27 Crabs: 12; Mantises: 21; Dragonflies: 11

Problem 8.28 $2: 60; $5: 30; $10: 30

Problem 8.29 Big mice: 12: Baby mice: 48

Problem 8.30 35

Problem 8.31 3

Problem 8.32 2

Problem 8.33 9

Problem 8.34 Highway: 4 miles; City Streets: 8 miles

Problem 8.35 5

Made in the USA
San Bernardino, CA
29 August 2018